FLAVONOIDS IN CELL FUNCTION

ADVANCES IN EXPERIMENTAL MEDICINE AND BIOLOGY

FLAVONOIDS IN CELL FUNCTION

Edited by

Béla S. Buslig

and

John A. Manthey

USDA Citrus and Subtropical Products Laboratory
Winter Haven, Florida

Kluwer Academic / Plenum Publishers
New York, Boston, Dordrecht, London, Moscow

Proceedings of the Symposium, Flavonoids in Cell Function, held March 29–30, 2000, during the 219th National Meeting of the American Chemical Society in San Francisco, California

ISBN 0-306-47254-6

©2002 Kluwer Academic/Plenum Publishers, New York
233 Spring Street, N.Y., NY 10013

http://www.wkap.nl/

10 9 8 7 6 5 4 3 2 1

A C.I.P. record for this book is available from the Library of Congress

Printed in the United States of America

PREFACE

The discovery of biological activity associated with flavonoid contaminants in vitamin C preparations from bell peppers and lemons by Szent-Györgyi and his associates opened a floodgate of research into the biological functions of this ubiquitous and diverse group of compounds. Since then, a broad range of physiological and biochemical activities were discovered in living systems including most plants and animals. With the continued discovery, isolation and identification of new natural and synthetic compounds exhibiting biological activities, entire research programs are devoted to wide ranging investigations to nearly every conceivable area, from microbial and plant interaction, growth regulation and development to physiological, genetical, medicinal actions and uses in animals.

This volume is based on presentations made at a Symposium, titled *Flavonoids in Cell Function*, held during the 219th National Meeting of the American Chemical Society held in San Francisco, California on March 29-30, 2000. The book is not intended to be a comprehensive treatise on flavonoid research, only a sampling of recent results. The papers cover a range of topics discussing various approaches to flavonoid study, starting at plant-microbe communication through analytical methods to medicinal and systemic implications of these compounds in animal cells and systems.

The organizers would like to express their thanks to Cargill Foods, Inc., Minneapolis, Minnesota and the Division of Agricultural and Food Chemistry of the American Chemical Society for financial support. A great deal of thanks is also due to the authors without whose cooperation and patience this volume would not be realized.

Béla S. Buslig and John A. Manthey

CONTENTS

FLAVONOIDS IN CELL FUNCTION

John A. Manthey[1], Béla S. Buslig[1] and Michael E. Baker[2]

Flavonoids (and isoflavonoids) are a large subgroup of secondary metabolites categorized as phenolics or polyphenols. Their wide distribution in the plant kingdom and prokaryotes, and the exceptional variety of these compounds, over 4000 have been identified so far, contributed to the attractiveness for study by chemists, geneticists, taxonomists and investigators in numerous other areas. The early investigators concentrated primarily on the purification and identification of these compounds. The studies of chemical structures of phenolics and their biochemical interrelations eventually expanded to cover the area of classification, leading to a distinct field of chemical or biochemical taxonomy. This approach greatly aided and often clarified classification schemes based on traditional methods relying on morphology, and later, genetic studies. The literature dealing with the chemistry, biochemistry and related aspects of flavonoids is extensive. This volume is not intended to be an exhaustive reference on flavonoids. Excellent information on these compounds can be found in numerous compilations of the chemistry, isolation and other techniques (Geissman,1962; Harborne, 1967; Mabry *et al.,* 1970; Harborne *et al.,* 1975; Harborne and Baxter, 1999).

The functions of these compounds in microorganisms, plants, plant-microbe communication, mammalian cell systems and techniques of isolation and identification are discussed in the following chapters.

1. PLANTS

The complexities of flavonoid function in both microorganisms and higher plants have been investigated extensively (see Manthey and Buslig, 1998; Winkel-Shirley, 2001). In the organization of this volume, the functional roles of flavonoids in microbe/plant interactions

[1] U.S. Department of Agriculture, Citrus and Subtropical Products Laboratory, 600 Avenue S, NW, Winter Haven, FL 33881; [2]Department of Medicine, 0623B, University of California, San Diego, 9500 Gilman Drive, La Jolla, CA 92093-0623.

Flavonoids in Cell Function, edited by Béla Buslig and John Manthey.
Kluwer Academic/Plenum Publishers, New York, 2002.

is followed through actions of these compounds in the development of higher plants. The topic first discussed by Straney *et al.* covers microbial recognition of host plants and the interactions between pathogenic fungi and these plants. Their work shows the functions of specific host flavonoids in the induction of developmental genes necessary for pathogenicity in *Nectria haematococca* MPVI. The interactions between the host plants and the microorganisms illustrate the complexity of the seemingly simple matter of pathogenesis. The complexity of the plant/microbe interactions and the ramifications of an interactive system modulated by flavonoids is illustrated by Vierheilig and Piché in their discussion of the symbiotic relationships exhibited by arbuscular mycorrhizal fungi (AMF). Extending earlier studies (Vierheilig *et al.* 1998), their work illuminates the difficulties encountered when dealing with multifaceted complex systems. As further advances are made into the nature of complex symbiotic interactions between host and guest organisms, it becomes apparent that systematic investigation frequently yields multiple answers, which lead to additional questions unanticipated at the beginning. In their discourse, Vierheilig and Piché draw attention to the problems encountered with assumptions based on similarities, such as between symbiosis with rhizobial and mycorrhizal systems, indicating that the conclusions reached with one system may not necessarily apply to the other. Although the results indicate that there is a definite involvement of flavonoids in the communication between the host plants and the AMF, and the answer may be tantalizingly close, a broader look at the mycorrhizal fungus, plant and soil microbial system interaction is warranted.

Progressing to higher plants, the work described by Taylor and Miller uses the requirement for flavonols for pollen germination and tube growth by employing a photo-activatable flavonol analog to probe the role of flavonol-3-*O*-galactosyltransferase in pollen germination. The flavonoid analog acts as an inhibitor of flavonol-induced pollen germination in a concentration dependent manner. The compound is an inhibitor of flavonol-3-*O*-galactosyltransferase, mediated by UV-A light treatment. This approach, utilizing the binding characteristics of affinity labeled flavonoid analog substrates/inhibitors can be used to identify the residues required for flavonol-binding and catalysis of enzyme systems in plant development.

In the contribution by Woo *et al.* extracts from wild type and antisense mutant alfalfa and *Arabidopsis* plants were analyzed by HPLC to examine the flavonoid components involved in plant development. The mutant plants are impaired in cell division and differentiation. In view of the involvement of flavonoids with the metabolism of indole-acetic acid (Furuya *et al.*, 1962; Galston, 1969), and auxin transport (Jacobs and Rubery, 1988), the significant quantitative differences observed in some of the flavonoids between wild type and mutant plants strongly suggest their role in plant development.

As plants serve as part of the diet for most of the animal kingdom, their components, flavonoids among them, are ingested and may interact with many of the constituents essential to life, such as enzymes and other receptors controlling growth and development. While the evidence for such interactions is based on both historical and modern observations (see the chapter by Riddle), the analysis of plant components is an important function, particularly in view of the increased awareness of the importance of phyto-chemicals in human health. The chapter by Berhow gives an overview of the various techniques useful in the analysis of flavonoids and related compounds. By necessity, the descriptions of the various techniques are relatively brief, as it was intended to be used with supplemental literature. However, the extensive bibliography covers all of the techniques mentioned quite adequately. Earlier works by Harborne (1967), Harborne *et al.* (1975) and Mabry *et al.* (1970) contain valuable compilations of classical techniques employed in flavonoid identification.

The extraordinarily large variety of flavonoids and related polyphenols, particularly isomeric or substituted species, which are indistinguishable by optical spectroscopy require techniques which are capable of distinguishing between closely related compounds. High performance liquid chromatography coupled with mass spectrometry (LC-MS) has recently became the method of choice (Barnes *et al.*, 1998; Cimino *et al.*, 1999; Holder *et al.*, 1999; Stevens *et al.*, 1999; Doerge *et al.*, 2000; Shelnutt *et al.*, 2000), primarily due to the increased sensitivity, versatility, specificity and in a large part to reduced cost. The chapter by Barnes *et al.* illustrates the use of LC-MS for the identification of specific isoflavonoids from the American groundnut, particularly in their conjugated forms. The chapter illustrates the optimization of methods for quantitative measurements of isoflavonoids, and covers approaches to substantially increase the sensitivity of the measurement.

2. ANIMALS

Although flavonoids were recognized for over a century, the report of the beneficial action of bell pepper or citrus flavonoids on capillary function by Rusznyák and Szent-Györgyi (1936) unleashed renewed interest in their biological activities. Evidence for medicinal use of plants and plant extracts can be found in the writings of numerous ancient civilizations. In the chapter by Riddle, he presents historical evidence for the use of plant material for the treatment of certain medical conditions based on historical evidence and the relationship to some recent significant medicinal discoveries. He draws attention to the practical value of examining historical records, folk medicine and anecdotal evidence to aid the discovery of new medicinal compounds, which may be essentially a rediscovery, or identification of the active ingredient in remedies employed by healers in the distant past.

In the last decades epidemiological studies have shown that diets rich in fruits and vegetables have important health benefits, including protection against heart disease, cancer, and numerous other diseases (Steinmetz, and Potter, 1991; Block *et al.*, 1992; Ames *et al.*, 1995). Although the details of how diet protects against disease are still being elucidated, it is clear that flavonoids have important roles in promoting health. The papers in this volume discuss new discoveries on mechanisms of flavonoid action that can lead to better use of fruits and vegetables in preventive medicine. Flavonoid action has traditionally been divided into two general mechanisms: anti-oxidant (Vinson, 1998; Hodnick *et al.*, 1998) activity and hormonal activity (Baker, 1998).

Some of the beneficial activities in animals of compounds in plants are likely to have a co-evolutionary basis because animals and plants have coexisted for hundreds of millions of years (Baker, 1998). During that time, animals have learned to use plants as a source of energy, vitamins and protection against disease. Flavonoids and other compounds such as vitamin E in plants have anti-oxidant activity that protects the plants from UV radiation and free radicals formed during photosynthesis. Plants and animals diverged from a common ancestor about 1.2 billion years ago. Many of the enzymes in plants and animals have retained sufficient similarity to respond to similar compounds.

One of the beneficial effects studied is the so called "French paradox", where it was observed that populations consuming moderate amounts of red wine and red or purple grape juice appear to acquire increased protection against vascular disease. In the chapter by Folts the potential health benefits of flavonoids found in grape juice are discussed. He concludes that a main effect may be due to the inhibition of the initiation of atherosclerosis by one or more mechanisms discussed in the paper. The existing evidence of antiplatelet and anti-oxidant benefits and improved endothelial function from red wine and purple grape juice suggests that moderate amounts of red wine or purple grape juice be included among the 5-7 daily servings of fruits and vegetables per day as recommended by the American Heart Association to help reduce the risk of developing cardiovascular disease.

Examining the antioxidant effects associated with citrus juices, Vinson *et al.* show quantitative evidence for the beneficial effects attributable to such activities of phenolic compounds, mainly flavonoids, *in vitro* and in some animal models *in vivo*. Accumulated evidence, building on earlier work (Vinson, 1998), indicates the importance of the quality of antioxidant activity, due to certain flavonoids and other polyphenols, in relation to heart disease.

In the work by Franke *et al.* the activity of two flavonoids, (+)-catechin and hesperidin is examined for their ability to prevent or reverse carcinogenesis in a colonic aberrant crypt (a form of precancerous cell) assay. Carcinogenesis is believed to be a multistep process involving repeated DNA damage, initiation, proliferation, clonal selection and progression (Arnold *et al.*, 1995). Inhibition of neoplastic transformation in cell cultures by potential chemopreventive agents, such as flavonoids, has been recommended due to relatively low costs, high speed, inclusion of the various stages of carcinogenesis and most importantly, due to the good prediction of *in vivo* activity. The results with catechin and hesperidin indicate the potential of these flavonoids to act as useful dietary chemopreventive agents in light of the good bioavailability of these agents in the circulation. However, their mechanism of action is not specified.

In the multistep process of carcinogenesis cell-cell interactions and the activity of immunocytes are very important. In the paper by Bracke *et al.* two effects of the citrus methoxyflavone, tangeretin, are investigated. Tangeretin is found to affect the function of the E-cadherin/catenin complex in human MCF-7/6 breast carcinoma cells, which results in firm cell-cell adhesion and inhibition of invasion *in vitro*. It also affects the functioning of the interleukin-2 receptor on T-lymphocytes and natural killer cells. This leads to a decrease in the cytotoxic competence of these immunocytes against cancer cells. At high oral doses, this second effect can abrogate the therapeutic suppression of tumor growth exerted by tamoxifen administered concurrently. No evidence for a tumor promoting effect of tangeretin by itself was found in these experiments. Results suggest that tangeretin may be applicable in cases where immunosuppression could be clinically beneficial.

Some of the activities observed with flavonoid compounds involve the interference with enzyme activity. Logically, due to the redox nature of most of these compounds, target enzymes are expected to involve oxidation/reduction reactions. In the work of Moini *et al.* a complex mixture of polyphenols obtained from French maritime pine bark (PBE) was investigated. Observations indicated that the inhibition by PBE of xanthine oxidase (XO) and xanthine dehydrogenase (XDH) was uncompetitive and showed a mixed mode action. Both PBE and XO are redox active. No correlation was observed between antioxidant activity and XO inhibition by PBE and other selected purified flavonoids. The results point to the importance of binding of PBE to XO in the enzyme inhibition rather than interference with the intramolecular electron transfer processes. Thereby, in addition to the redox based effects, PBE has direct protein binding properties, adding another biochemical basis of its mode of action in biological systems.

Hormonal activities have focused on binding of flavonoids to steroid receptors or to enzymes involved in steroid metabolism due to structural similarities between flavonoids and steroids. The best studied interaction has been binding of flavonoids to the estrogen receptor. Some flavonoids have estrogenic activity; other flavonoids bind to the estrogen receptor, but do not activate transcription of estrogen-responsive genes. However, it is clear that flavonoids can affect steroid hormone action by binding to other sites, such as enzymes important in steroid hormone synthesis or degradation and proteins involved in transport of steroids to target cells. Moreover, many other proteins are important in steroid hormone action. Binding of flavonoids can have important hormonal actions (Baker, 1998).

Phytoestrogens present in the diet are supposed to have beneficial effects on the development and progression of a variety of endocrine-related cancers. Krazeisen *et al.* have examined the effect of a variety of dietary phytoestrogens, especially flavonoids, on the activity of human 17β-hydroxysteroid dehydrogenase type 5 (17β-HSD 5), a key enzyme in the metabolism of estrogens and androgens. These studies show that the reductive and oxidative activities of the enzyme are inhibited by many compounds, especially zearalenone, coumestrol, quercetin and biochanin A. Among flavones, inhibitor potency is enhanced with increased degree of hydroxylation. The most effective inhibitors seem to bind to a hydrophilic cofactor binding pocket of the enzyme. Schemes are presented to show the inhibitory strength of the various flavonoids for oxidative and reductive reactions catalyzed by this enzyme.

Flavonoids show various effects on animal cells, such as inhibition of platelet binding and aggregation, inhibition of inflammation, and anticancer properties, but the mechanisms of these effects remain largely unexplained. The interactions between these compounds and adenosine receptors (which are involved in the homeostasis of the immune, cardiovascular, and central nervous systems) and adenosine agonists/antagonists which exert many similar effects is the topic of the chapter by Jacobson *et al.* The affinity of flavonoids and other phytochemicals to adenosine receptors suggests that a wide range of natural substances in the diet may potentially block the effects of endogenous adenosine. Competitive radioligand binding assays were employed to screen flavonoid libraries for affinity for the receptors, along with comparative molecular field analysis calculations to establish steric and electrostatic requirements for ligand recognition at three subtypes of adenosine receptors. Adenosine receptor antagonism is suggested to be involved in biological activities reported for the flavonoids.

One of the serious health problems affecting the American population is coronary heart disease (CHD), associated with elevated cholesterol levels. High cholesterol levels, and especially low-density lipoprotein (LDL) cholesterol is thought to be critical in the formation and development of atherosclerotic plaque, the underlying pathological condition of CHD. Epidemiological studies indicate that high dietary intake of fruits and vegetables reduced the risk of CHD development. The cardioprotective effect of the dietary flavonoids is thought to be related to the antioxidant or anti-aggregant properties of these compounds. The chapter by Kurowska and Manthey, shows the effect of various concentrations of polymethoxyflavones on the reduction of the secretion of the structural protein (*apoB*) present in low density lipoprotein. These results are encouraging, but no mechanism for the effect is proposed.

Restoration of blood flow to ischemic tissue initiates a complex series of deleterious reactions, ultimately the same effects as ischemia, i.e., cell injury and necrosis. Recent work indicates that flavonoids are particularly effective anti-inflammatory agents in the setting of ischemia/reperfusion (I/R). Korthuis and Gute address the protective effect observed upon administration of flavonoids. While the underlying mechanisms of the protective effects of these compounds is uncertain, evidence indicates that flavonoids are potent anti-oxidants which also inhibit key regulatory enzymes involved in the activation of pro-inflammatory signaling cascades. In addition, it appears that these compounds prevent the expression of specific adhesion molecules involved in leukocyte recruitment, observations which provide the molecular basis for the anti-adhesive properties of these compounds. To limit the post-ischemic inflammatory response, administration of a micronized mixture of 90% diosmin and 10% hesperidin (Daflon 500 mg) is particularly efficacious. The potent anti-oxidant nature of these compounds along with inhibition of key regulatory enzymes involved in the generation of powerful proinflammatory signaling cascades appear to be essential for their action. Flavonoids also appear to prevent the expression of specific adhesion molecules involved in leukocyte recruitment, which may be a molecular basis for the anti-adhesive effects of these compounds.

The broad range of flavonoids and their effects on a variety of physiological and developmental realms includes the regulation of the expression of many genes. In the final chapter, Kuo tackles the complex subject of cellular regulation by several flavonoids. Little information exists regarding the contribution of various structural elements or on their molecular mechanisms of action. The explanation for stimulatory or inhibitory effects on gene expression may lie in the estrogenic behavior of certain flavonoids. The affinity of flavonoids for estrogen receptors (ER) could explain the stimulatory effect on genes with estrogen-responsive elements (ERE). However, other modes of action apparently also exist and need to be further explored. The physiological relevance of the regulation of gene expression by environmental chemicals such as flavonoids is always problematic. One factor of concern is the *in vivo* concentration of flavonoids. Besides intestinal cells, liver cells and skin cells, other tissues obtain flavonoids through blood circulation. Thus, plasma concentrations of flavonoids are normally discussed. Compared with the concentrations of flavonoids used in cell culture systems to demonstrate their effectiveness, the plasma concentrations are low. Nevertheless, evidence shows that some flavonoids accumulate in cells. Intracellular transport and the accumulation of these compounds need to be reconciled to explain the differences in results between *in vitro* cell culture and systemically applied flavonoids. If intracellular accumulation of certain flavonoids is a shared characteristic for various cell types, it implies that routine ingestion of flavonoids could lead to biological effects at concentrations lower than predicted from a single treatment. Further experiments to address possible cell/tissue accumulation of flavonoids are greatly needed.

3. REFERENCES

Ames, B. N., Gold, L. S., and Willet, W. C., 1995, The causes and prevention of cancer. Proc. Natl. Acad. Sci. USA **92**, 5258-5265.

Arnold, J. T., Wilkinson, B. P., Sharma, S., and Steele, V. E., 1995, Evaluation of chemopreventive agents in different mechanistic classes using a rat tracheal epithelial cell culture transformation assay, *Cancer Research* **55**:537-543.

Baker, M. E., 1998, Flavonoids as hormones: A perspective from the analysis of molecular fossils, In: *Flavonoids in the Living System*, Manthey, J. A. and Buslig, B. S., eds., Plenum Press, New York and London, pp. 249-267.

Barnes, K. A., Smith, R.A., Williams, K., Damant, A.P., and Shepherd, M. J., 1998a, A microbore high performance liquid chromatography/electrospray ionization mass spectrometry method for the determination of the phytoestrogens genistein and daidzein in comminuted baby foods and soya flour, *Rapid Commun. Mass Spectrom.* **12**:130-138.

Block, G., Patterson, B., and Subar, A., 1992, Fruit, vegetables and cancer prevention: a review of the epidemiologic evidence. Nutr. Cancer **18**, 1-29.

Cimino, C. O., Shelnutt, S. R., Ronis, M. J. J., and Badger, T. M., 1999, An LC-MS method to determine concentrations of isoflavones and their sulfate and glucuronide conjugates in urine, *Clin. Chim. Acta* **287**:69-82.

Doerge, D. R., Chang, H. C., Churchwell, M. I., and Holder, C. L., 2000, Analysis of soy isoflavone conjugation *in vitro* and in human blood using liquid chromatography-mass spectrometry, *Drug Metab. Disp.* **28**:298-307.

Furuya, M., Galston, A. W., and Stowe, B. B., 1962, Isolation from peas of cofactors and inhibitors of indolyl-3-acetic acid oxidase, *Nature* **193**:456-457.

Galston, A. W., 1969, Flavonoids and photomorphogenesis in peas, In *Perspectives in Phytochemistry*, Harborne, J. B., and Swain, T., eds., Academic Press, New York., pp. 193-204.

Geissman, T. A., 1962, *The Chemistry of Flavonoid Compounds*, Macmillan, New York.

Harborne, J. B., 1967, *Comparative Biochemistry of the Flavonoids*, Academic Press, New York.

Harborne, J. B., and Baxter, H., eds., 1999, *The Handbook of Natural Flavonoids*, 2 vols., CHIPS Books, 1800 pp.

Harborne, J. B., Mabry, T. J., and Mabry, H., 1975, *The Flavonoids*, Chapman and Hall, London.

Hodnick, W. F., Ahmad, S., and Pardini, R. S., 1998, Induction of oxidative stress by redox active flavonoids, In: *Flavonoids in the Living System*, Manthey, J. A. and Buslig, B. S., eds., Plenum Press, New York and London, pp. 131-150.

Holder, C. L., Churchwell, M. I., and Doerge, D. R., 1999, Quantification of soy isoflavones, genistein and daidzein, and conjugates in rat blood using LC/ES-MS, *J. Agric. Food Chem.* **47**:3764-3770.

Jacobs, M., and Rubery, P. H., 1988, Naturally occurring auxin transport regulators, *Science* **241**:346-349.

Mabry, T. J., Markham, K. R., and Thomas, M. B., 1970, *The Systematic Identification of Flavonoids*, Springer-Verlag, New York.

Manthey, J. A., and Buslig, B. S., eds., 1998, *Flavonoids in the Living System*, Plenum Press, New York and London.

Rusznyák, S. and Szent-Györgyi, A., 1936, Vitamin P: flavonols as vitamins, *Nature* **27**:138.

Shelnutt, S. R., Cimino, C. O., Wiggins, P. A, and Badger, T. M., 2000, Urinary pharmacokinetics of the glucuronide and sulfate conjugates of genistein and daidzein, *Cancer Epidem. Biomark. Prevent.* **9**:413-419.

Steinmetz, K. A., and Potter, J. D., 1991, Vegetables, fruit, and cancer. I: Epidemiology. Cancers Causes Control **2**, 325-357.

Stevens, J. F., Taylor, A.W., and Deinzer, M.L., 1999, Quantitative analysis of xanthohumol and related prenyl-flavonoids in hops and beer by liquid chromatography-tandem mass spectrometry, *J. Chromatogr. A* **832**:97-107.

Vierheilig, H., Bago, B., Albrecht, C., Poulin, M.-J., and Piché, Y., 1998, Flavonoids and arbuscular-mycorrhizal fungi, In: *Flavonoids in the Living System*, Manthey, J. A. and Buslig, B. S., eds., Plenum Press, New York and London, pp. 9-33.

Vinson, J. A., 1998, Flavonoids in foods as *in vitro* and *in vivo* antioxidants, In: *Flavonoids in the Living System*, Manthey, J. A. and Buslig, B. S., eds., Plenum Press, New York and London, pp. 151-164.

Winkel-Shirley, B., 2001, Flavonoid biosynthesis. A colorful model for genetics, biochemistry, cell biology, and biotechnology, *Plant Physiol.* **126**:485-493.

HOST RECOGNITION BY PATHOGENIC FUNGI THROUGH PLANT FLAVONOIDS

David Straney[1], Rana Khan, Reynold Tan and Savita Bagga

1. ABSTRACT

A common characteristic among fungal pathogens of plants is that each specializes on a narrow range of specific plants as hosts. One adaptation to a specific host plant is the recognition of the host's chemicals which can be used to trigger genes or developmental pathways needed for pathogenesis. The production of characteristic flavonoids by plants, particularly those exuded from roots by legumes, appear to be used as signals for various microbes, including symbionts as well as pathogens. *Nectria haematococca* MPVI (anamorph: *Fusarium solani)* is a soil-borne pathogen of garden pea (*Pisum sativum*) which serves as a useful model in studying host flavonoid recognition. This fungus displays flavonoid induction of specific pathogenicity genes as well as stimulation of development needed for pathogenesis. Here, we summarize the study of flavonoid-inducible signal pathways which regulate these trait, through identification of transcription factors and regulatory components which control these responses. The characterization of the components a pathogen uses to specifically recognize its host provides insights into the host adaptation process at the molecular level.

2. PATHOGEN ADAPTATION TO HOST FLAVONOIDS

Plants produce flavonoids for a variety of reasons, including UV protectants, pigments, and hormones (Bohm, 1998). Flavonoids also play a significant role in how the plants interacts with other organisms in the environment. Flavonoids produced by a number of plants appear to play a defensive role, providing anti-feeding activity against insect pests or possessing fungistatic activity for protection from pathogenic fungi, such as the stress-

---

[1] University of Maryland, Department of Cell Biology and Molecular Genetics, College Park, MD 20742-5815

Flavonoids in Cell Function, edited by Béla Buslig and John Manthey.
Kluwer Academic/Plenum Publishers, New York, 2002.

                                                           D. STRANEY *ET AL.*

induced phytoalexins. These two protective activities are shared among a number of legume isoflavonoid phytoalexins (Lane et al., 1987). A hallmark of plant secondary metabolites is the diversity observed among different plants, and this is particularly true of the flavonoids (Harborne, 1971).

One important consequence of this diversity is that a pathogen must adapt to tolerate those defense flavonoids produced by its host. This may be one factor which limits the host range of any one fungal strain, due to a difficulty in evolving resistance to multiple defense compounds (VanEtten et al., 1989), or result from a stepwise adaptation to greater pathogenesis on the specialized host. Another consequence of the chemical diversity of plants is that the host chemicals may be used by the pathogen to detect its host, necessary for a highly specialized pathogen. These fungal systems provide a system to examine a range of examples of co-evolution of the host and pathogen. The microbial nature of fungi and their mode of interaction with plants at the cellular level allows a focus upon relatively simple and defined responses and identification of molecular components which control these responses.

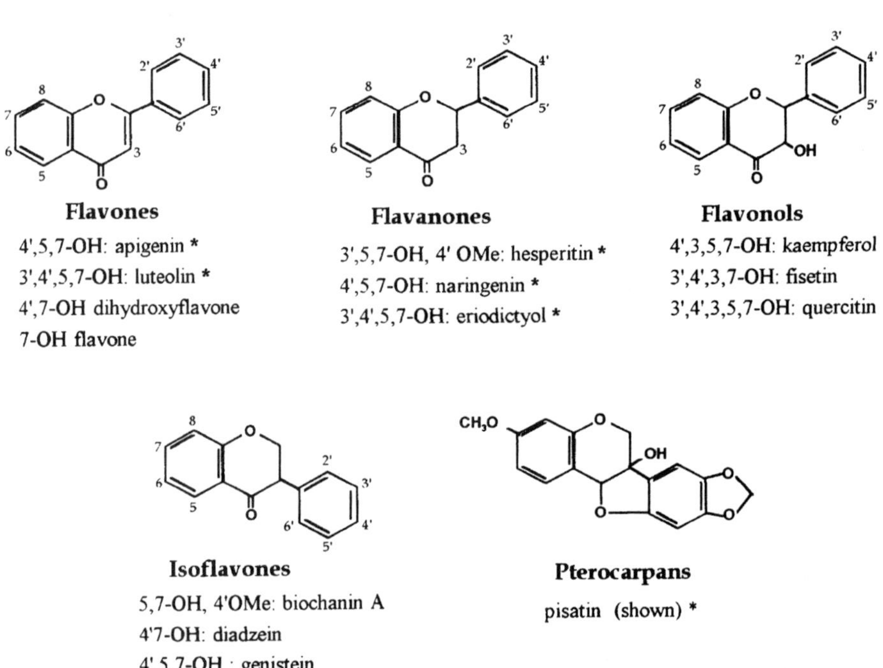

Figure 1. Structures of flavonoids. The major classes of flavonoids discussed are shown with examples of specific flavonoids (position of -OH or -OMe substitution). Pisatin is the phytoalexin produced by pea. The detoxification of pisatin by pisatin demethylase removes the 7-*O*-methyl group. All compounds shown have been tested for ability to stimulate germination of *Fusarium solani f sp. pisi* germination and those marked (*) produce greater than 50% germination at 50 µM flavonoid 'in succinate medium. Under similar conditions, succinate medium alone produces 10% germination. Data from Ruan & Straney 1995.

One form of pathogen response to host defense flavonoids is to develop a tolerance to those compounds. A survey among a number of fungi demonstrated that only those pathogenic on pea were able to grow in the presence of pisatin, a fungistatic isoflavonoid, produced by this plant (Fig. 1) (VanEtten, 1973). This same general result is seen among other plant-fungal relationships (VanEtten et al., 1989). Degradation of the host defense compound is one sort of tolerance, characterized with a number of legume isoflavonoid phytoalexins (VanEtten et al., 1989). Several genes encoding degradative enzymes have been cloned from *Fusarium solani* strains which specialize upon different legumes, including a pisatin demethylase from a pea pathogen (Maloney and VanEtten, 1994), maakiain hydroxylase from a chickpea pathogen (Covert et al., 1996) and kievitone hydrolase from a bean pathogen (Li et al., 1995). Each enzyme displays a high degree of specificity to the phytoalexin of its host, implying a specific adaptation to each host's defense compound. Similar degradative systems exist in pathogens of other plants, specific for saponin, cyanogenic and sesquiterpene defense compounds of their hosts (Morrissey and Osbourne, 1999). In addition to degradative tolerance, some fungi also demonstrate a non-degradative tolerance to host compounds. The pea-pathogenic *F. solani* shows preferential export of the pea phytoalexin pisatin (Denny et al., 1987). The action of ABC transporters in fungi (Del Sorbo et al., 1997) suggests that specific isoflavonoid phytoalexin export proteins may be easily identified.

2.1 Adaptation To Recognize Flavonoids As Host-Specific Signals

Beyond developing means to just tolerate a host flavonoid, there are indications that microorganisms can utilize chemicals produced by their host plants to coordinate development and gene expression during pathogenesis. Such host recognition has been particularly well characterized in pathogenic bacteria (van Gijsegem, 1997). Examples include induction of hypersensitivity and pathogenicity *(hrp)* gene clusters by nutritional cues encountered in the host (Wengelnik et al., 1996; Xiao et al., 1992), induction of a *Pseudomonas syringae* toxin, syringomycin, by plant phenolic-glycosides (Mo et al., 1995), and induction of the *vir* genes in *Agrobacterium* by phenolics and sugars encountered in wounds of host plants (Winans, 1992). The most specific host cues used among bacteria-plant interactions is that in the nitrogen fixing rhizobial symbionts of legumes. Flavonoids exuded from legume roots and seed coats are recognized by rhizobia and used to induce bacterial *nod* gene expression required for subsequent events in nodulation (Pueppke, 1996). Among the *rhizobium* strains which specialize on different legumes, the *nod* gene inducers match those flavonoids exuded by the host plant (Pueppke, 1996). Host flavonoids also induce other responses, independent of *nod* gene signaling, which include growth promotion, chemotaxis and secretion of specific proteins (Dharmatilake and Bauer, 1992; Hartwig et al., 1991; Krishnan and Pueppke, 1993). Although these processes are not the sole determinants of rhizobial host specificity, the ability to recognize flavonoids released by their respective hosts appears to be a critical first recognition step in the *rhizobium*-legume interaction. It would seem unlikely that the flavonoids exuded into the rhizosphere would not be used by other microbes which interact with legumes. In fact, a number of fungi do respond to flavonoids produced by their host. Chemotaxis of motile zoospores to host flavonoids has been described for *Aphanomyces euteiches* (Yokosawa et al., 1986), a pea pathogen, and *Phytophthora sojae*, a soybean pathogen (Morris and Ward, 1992; Tyler et al., 1996). Other soybean isoflavonoids were found to induce encystment and germination of *Phytophthora sojae* (Morris and Ward, 1992). Flavonoids have been reported to stimulate germination of zygospore of vesicular-arbuscular (VA) endomycorrhizal *Gigaspora* and *Glomus* strains which interact with legumes (Gianinazzi-Pearson et al., 1989; Tsai and Phillips, 1991). Progress has been made in determining the mechanisms for flavonoid recognition and response in these fungi. Similarities of flavonoid and estrogen-binding sites have been noted in the endomycorrhyzae which germinate in response to flavonoids (Poulin et al., 1997), and calcium signaling has been implicated in *Phytophthora* zoospore encystment (Connolly et al., 1999).

3. *FUSARIUM SOLANI* (*Nectria haematococca* MPVI) AS A SYSTEM FOR STUDYING RESPONSE TO HOST FLAVONOIDS

Fusarium solani is a soil borne plant pathogen. Although it causes root- and foot-rot disease on a number of hosts, including pea, bean and soybean, different strains show a strong host specificity. These host specialization groups appear to be genetically distinct (O'Donnell and Gray, 1995). The pea pathogenic strain is particularly valuable for the study of gene action since it belongs to a sexual state (*Nectria haematococca* MPVI), allowing genetic analysis. Further DNA transformation techniques have been developed to allow manipulation and disruption of genes (VanEtten and Kistler, 1988). As such, it serves as a model system for the other *F. solani* legume interactions. It may additionally model the host adaptation of the wilts affecting many legumes, caused by different host-specific strains of *Fusarium oxysporum*. In keeping with fungal taxonomic preferences, and in describing the genetics of the pea pathogenicity, the pea-pathogenic *F. solani* will be referred to as its sexual state *Nectria haematococca* MPVI, which includes strains which lack pathogenicity on pea. This fungus displays a number of traits which are regulated by host flavonoids, both at the gene level, and in its development. These will be discussed separately in the following sections.

3.1. The *PDA1* Gene as a Model for Host-Isoflavonoid Regulated Genes

Pea produces the isoflavonoid pterocarpan pisatin (Fig. 1). The antifungal activity of pisatin and its induced production after abiotic stress or microbial challenge, earn its classification as a phytoalexin. Other legumes produce similar pterocarpans as phytoalexins, such as glyceollin (soybean) and phaseollin (bean), but this isomer of pisatin is restricted to *Pisum* and its close relative *Lathyrus*. An ability to tolerate pisatin is common to many pea pathogens, and a degradative activity, pisatin demethylase, has been characterized in pea pathogenic strains of *N. haematococca*, *Fusarium oxysporum*, *Ascochyta pisi*, and *Mycosphaerella pioides* (George, 1993). The *N. haematococca* pisatin demethylase enzymatic activity has been characterized as a cytochrome P450 monooxygenase (George et al., 1998), an activity supported by conserved P450 sequence motifs found in a cloned *PDA* gene encoding pisatin demethylase (Maloney and VanEtten, 1994). Manipulation of the *PDA* gene show it to play a partial role in determining the virulence upon pea (Ciuffetti and VanEtten, 1996; Oeser and Yoder, 1994; Schäfer et al., 1989; Wasmann and VanEtten, 1996). The *PDA1* gene, resides upon a small conditionally dispensable 1.6 Mb chromosome (Miao et al., 1991). Loss of this chromosome does not affect growth in culture, but does severely attenuate virulence upon pea. This suggests that the chromosome is habitat specific, needed for growth on pea but not in other conditions (VanEtten et al., 1994). Recent experiments by Drs. Kistler and VanEtten have identified three additional genes necessary for pathogenesis of pea on the same 1.6 Mb chromosome that are in fact clustered next to the *PDA1* gene (Kistler et al., 1996).

What is particularly significant about the *PDA1* gene in terms of signaling is the regulation of its expression. Treatment of mycelia with pisatin causes a 20-fold increase in both *PDA* enzyme activity and *PDA1* mRNA. This induction is highly specific to pisatin (VanEtten et al., 1989). Thus, the fungus is able to recognize the presence of its hosts phytoalexin and respond by stimulating expression of the *PDA1* gene. Similar pisatin induction is also displayed by the pisatin demethylase activities characterized in other fungi pathogenic on pea (George et al., 1998). This regulation of the *PDA1* gene appears to be significant for pathogenesis since comparison of different forms of the *PDA* gene demonstrated that only the highly pisatin-inducible forms are consistently found in *N. haematococca* isolated from diseased pea (VanEtten et al., 1994). Further, pisatin induction is not limited to the *PDA1* gene. A non-degradative tolerance to pisatin, acting at the export of pisatin from mycelia, is also induced by pisatin (Denny et al., 1987). Additionally, preliminary evidence indicates that the expression of other genes located in the pea pathogenicity gene cluster with *PDA1* are induced by pisatin (VanEtten, unpublished). This broad regulation suggests that pisatin is utilized as a host-specific recognition signal, used to trigger expression of a number of genes required for pathogenesis of pea. Although there are many examples in pathogenesis where a host-specific signal appears to determine expression of key genes, these in *N. haematococca* represents the only case where a specific plant signal and target genes have been identified and allow their molecular characterization. This would allow comparison to fungal recognition of non-host specific signals from the plants, such as nutrient limitation and plant cell wall constituents (Guo et al., 1995; Hedge and Kolattukudy, 1997; Lin and Kolattukudy, 1978), which control other pathogenicity genes.

The *PDA1* gene offers a model to study how *N. haematococca* adapted to pea by developing a means to recognize pisatin and regulate gene expression. The approach towards identifying components in the pisatin- signaling pathway has focused upon the DNA elements in the *PDA1* promoter which provide pisatin- responsive transcription. Reconstitution of transcription from the *PDA1* promoter provided a means to easily modify sequence elements to determine their function. The in vitro techniques were successful in identifying a protein which bound in a pisatin-dependent manner and functioned as a transcriptional activator (Ruan and Straney, 1996; Straney et al., 1994; Straney and VanEtten, 1994). However, when in vivo expression was tested through transformation of *PDA1* promoter constructs fused to the GUS reporter gene, the putative pisatin-responsive element was found to not confer pisatin-responsive transcription (Khan and Straney, 1999). Promoter analysis using the reporter gene-based in vivo expression system has identified a 40 bp element which confers pisatin-inducible expression to a minimal promoter (Khan and Straney, unpublished). Pisatin control of transcription appears to act through a transcription factor which binds this DNA sequence. Identification of the sequence will allows the cloning of this transcription factor. Manipulation of this transcription factor through gene disruption will provide a means to determine if it controls all pisatin-induced genes in *N. haematococca*. Further, it will then be possible to determine whether the same transcription factor is conserved in other pea pathogens which display pisatin-induced pisatin demethylase activity, or if host recognition components evolved independently.

An important question in understanding the role of signaling in host-pathogen interactions is whether a signal which controls gene expression in culture also functions during pathogenesis. Experiments using cultured mycelia and purified chemicals can show an effect upon gene expression, but that compound may not reach relevant levels in the plant. *PDA1*, for example is induced during pathogenesis, concomitant with the onset of pisatin synthesis by the host (Hirschi and VanEtten, 1996; Khan and Straney, 1999). However, *PDA1* is regulated by other compounds besides pisatin. Its expression is partially induced by starvation of the mycelia for carbon (glucose) or nitrogen (amino acids) sources (Khan and Straney, 1999; Straney and VanEtten, 1994). This nutritional repression is shared by a number of pathogenicity genes, either for nitrogen sources (Larson and Nuss, 1994, Talbot et al., 1993; van den Ackerveken et al., 1994), or carbon sources (Gonzalez-Candelas and Kolattukudy, 1992; Lin and Kolattukudy, 1978; Pieterse et al., 1994).

Nutritional derepression has been suggested to be a common trigger of fungal pathogenicity genes, sensing a nutritionally-poor environment or on the outside surface of or within of the plant (Talbot et al., 1997). The dissection of control elements in the *PDA1* promoter provided a means to genetically manipulate the regulation of *PDA1* and so address the relative biological importance of pisatin and nutrient regulation during pathogenesis. Removal of nutritional control elements from the *PDA1* promoter did not significantly alter its induction during pathogenesis. Although nutrient levels might be high enough to alter expression during pathogenesis, they do not appear to modulate *PDA1* expression. In contrast, removal of pisatin control elements eliminated the induced expression (Khan and Straney, 1999). Thus, pisatin appears to provide a host-specific regulation of this gene.

3.2. Flavonoid Induction of Development

As discussed above, several soil-borne fungi which interact with legumes regulate certain stages of their development in response to flavonoids exuded from their hosts. These developmental transitions (zoospore encystment, spore germination; Fig. 2) are required for further growth and penetration of the plant. Although development involves gene regulation, it differs from that described for the *PDA1* gene in that large sets of genes are affected, and these may need to be regulated by other signals under different growth conditions. Thus, signal pathways which control sets of developmental genes are often more important targets of study rather than any one gene itself The control of germination appears to be particularly important in soil-borne fungi since propagules, such as spores, remain dispersed in the soil for months or years in a quiescent state until the appearance of a potential host stimulates germination. The propagules presumably recognize the presence of a plant through plant-derived soluble or volatile compounds since exudates from germinating seeds and roots stimulate germination of spores or sclerotia from a number of genera, including *Pythium, Phytophthora, Verticillium, Aphanomyces* and *Fusarium, Rhizoctonia* and *Sclerotium* (Nelson, 1991). This chemical recognition of the host is common among a number of other plant pathogens, including parasitic plants (Lynn and Chang, 1990) and nematodes (Gheysen, 1998). A classic example of fungal recognition of a host plant is the selective germination of *Sclerotium cepivorum* sclerotia towards its host *Allium* species. Germination of the pathogen is induced by volatile alkyl sulphides,

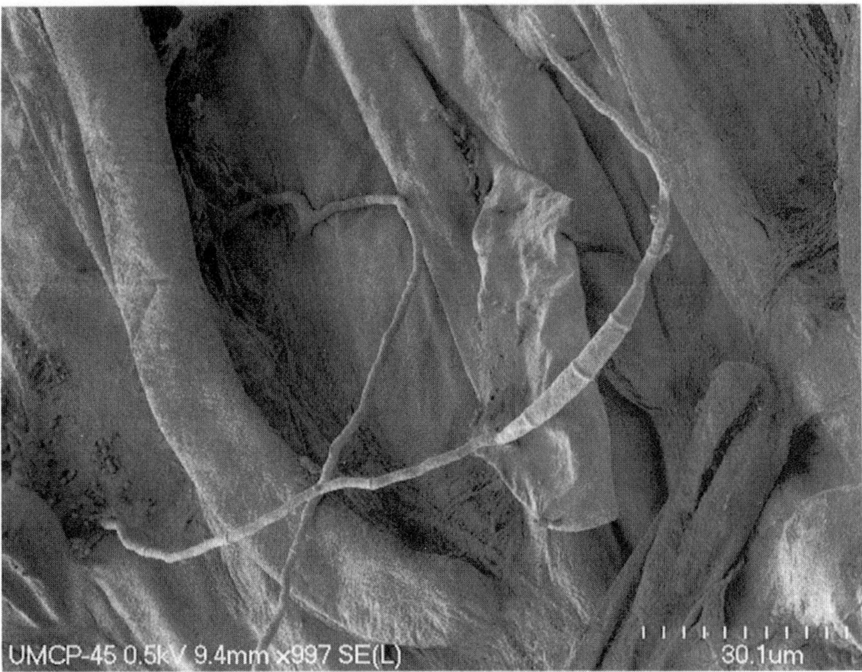

Figure 2. *Nectria haematococca* MPVI (*Fusarium solani*) macroconidia germinating on the surface of pea root. The spore is a four-celled structure. During germination, hyphal germ tubes emerge from the ends of the spore.

produced by bacterial metabolism of the water soluble propyl and allyl cysteines and sulphoxides released from *Allium* roots (Coley-Smith and King, 1970). Despite such examples of host-specific signaling, a dogma developed that nutrients leached from the roots trigger pathogen germination. Current research is proving this dogma to be too simplistic and is identifying different compounds acting as signals in other host-pathogen pairs which may explain the host selectivity of germination (Nelson, 1991).

Certain flavonoids induce germination of *N. haematococca* MPVI asexual spores, termed macroconidia and chlamydospores, which form the soil-borne inoculum of *Fusarium* species. Although the specific flavonoids exuded by pea have not been definitively identified, the overall flavonoid specificity of germination match that of *nod* gene induction in the pea-specific rhizobia, *Rhizobium leguminosarum* bv. *viciae* (Ruan et al., 1995). Two compounds tentatively identified in pea root exudates, apigenin-7-*O*-glucoside and erio-dictyol (Firmin et al., 1986), stimulate germination (Fig. 1). The host selectivity of germi-nation provided by flavonoids is consistent with that seen with root exudates. When exudates from different plants were compared for germination stimulating activity (Fig. 3A), pea root exudate displayed the highest activity. Another legume (soybean) and non-legumes did not produce stimulation of germination. Bean root exudate provided partial stimulation, however this may be expected from bean's exudation of eriodictyol and naringenin (Hungria et al., 1992), both germination stimulators. The levels of nutrients

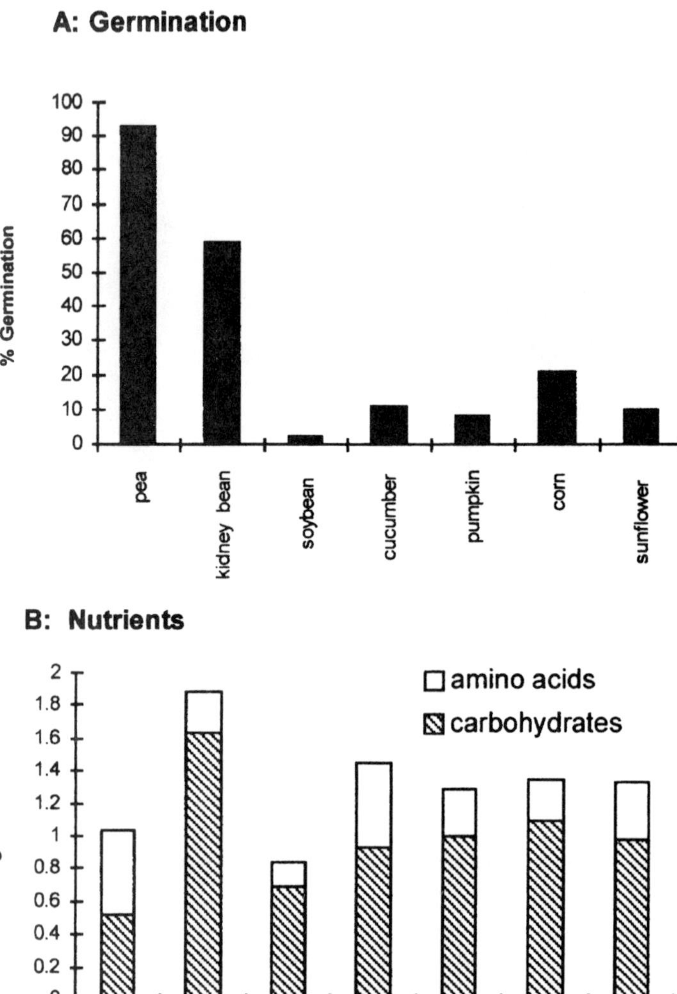

Figure 3. Host specificity of spore germination of *N. haematococca* strain 77-2-3. Root exudates were collected from seedlings between 4 and 8 days-old. *A*: Germination of macroconidia of *F. solani* f sp. *pisi*. The root exudates were either diluted or concentrated so that the amount of exudate used for each plant would represent the same weight of roots as the pea sample. Germination assays are compared with addition of 0.15x succinate medium (+ succ. medium) of no added medium (- succ. medium). Controls with no plant exudate produced 3% germination (+ succ.) and 0.5% germination (succ.). Mixture of pea exudate and non-inducing exudates displayed the same germination as the pea exudate. *B*: Nutrient levels measured in root exudates. Carbohydrates were measured as aldoses (hexoses and pentoses) after acid hydrolysis of oligomers and cysteine-sulfuric acid assay against a glucose standard. Amino acids were measured by measurement of their reaction with the fluorescent probe *o*-phthalaldehyde against a glycine standard. These amounts are also normalized to root weight.

(sugars, amino acids) in the root exudates did not correlate with the germination levels (Fig. 3B), suggesting that the higher germination in pea root exudates is not from higher nutrient levels. The results suggest that flavonoids exuded by pea into the rhizosphere to induce rhizobial nodulation may also be used by this pathogenic fungus as a host selective cue as well. Such host recognition by a pathogen does not appear to be an isolated case in *N. haematococca. Fusarium oxysporum* f. sp. *pisi*, another pathogen of pea displays a similar flavonoid specific germination as *N. haematococca*, and a bean pathogen, *F. solani* f. sp. *phaseoli*, displayed a different spectrum of flavonoid response more closely matching that exuded by bean (Ruan et al., 1995).

In addition to *nod*-inducing flavonoids, pisatin, the pea-specific pterocarpan phytoalexin, also strongly induced germination of *N. haematococca* spores (Ruan et al., 1995). Although phytoalexins are normally thought of as internal defenses, other pterocarpan phytoalexins have been shown to be exuded from soybean and alfalfa roots (Dakora et al., 1993; Graham, 1991; Schmidt et al., 1992). Phillips and Kapulnik (1995) proposed that legumes exude these phytoalexins to influence the composition of microbial populations in the rhizosphere. Pisatin is present in root exudates from peas grown under non-aseptic conditions and attains levels that would be highly stimulatory for germination. (R. Tan and D. Straney, unpublished). The greater host-specificity among pterocarpan phytoalexins than the *nod* inducing flavonoids present another host-selective signal for *N. haematococca* recognition of its host. Because plants produce phytoalexins in response to environmental stress or microbial challenge, these compounds may provide an additional strong host-specific stimulus from roots experiencing such stresses under natural conditions. Although the stimulatory activity of pisatin would seem to be contrary to that expected for phytoalexins, which are fungistatic, the pisatin degrading activity and non-degradative tolerance in *N. haematococca*, discussed above, would provide protection upon which further adaptation would be possible (VanEtten et al., 1989). The idea that a plant defense chemical could be utilized as a stimulatory signal is not new. Various monophagous insects which have come to tolerate a flavonoid antifeeding compound have been shown to use it as an attractant (Harborne and Grayer, 1993), However, this would be the first case of such in a fungal-plant interaction. The microbial nature of the fungal system provides tools for understanding how such host adaptation may occur.

The correlation of exudation of a compound and the ability of that pure compound to elicit a response is suggestive of a function, however, it is necessary to prove such action and determine its biological relevance in relation to other signals. Well characterized plant-bacterial interactions demonstrate that multiple signals may be acting at the same time, such as the action of *nod* inducing betaines as well as flavonoids in the *rhizobium*-legume interaction (Phillips et al., 1992) or wound phenolics and sugars in the *Agrobacterium*-host interaction (Winans, 1992). The contribution of any one signal may be relatively minor, or may vary depending upon environmental conditions. One approach to determine the relative function of flavonoids in *N. haematococca* spore germination is to fractionate pea exudates and determine the relative activity from different fractions. Fractionation of pea root exudates by C-18 reverse-phase HPLC has produced approximately three peaks of activity. One peak co-elutes with pisatin and displays a UV-spectrum characteristic of pisatin. This pisatin-containing fraction represents approximately one-half to three-quarters of the overall eluting germination-stimulating activity; this contribution is greater under conditions where the pea root is challenged with pea-specific rhizobia or a low-virulence fungal isolate where more pisatin is exuded (R. Tan and D. Straney, unpublished). Another earlier eluting peak which induces germination also induces the *nod* genes in *R. leguminosarum* bv *viciae* and so may contain *nod*-inducing flavonoids. This approach provides direct evidence for a dominant role of flavonoid signaling in this interaction. However, it is subject to artifacts from microbial or plant metabolism of flavonoids or other metabolites which may act under natural conditions. A molecular approach which identifies signal transduction pathways in the fungus used in flavonoid recognition would provide a means to specifically genetically alter the fungal response and so determine the effect upon the response under natural conditions.

Study of flavonoid-induced stimulation of *N. haematococca* germination has implicated the cAMP pathway as that controlling the germination response. Treatment of spores with cAMP induced germination, and treatment with H-89, an inhibitor of cAMP-dependent protein kinase A (PKA), prevented flavonoid-induced germination (Ruan et al., 1995). Further, treatment of spores with flavonoids produced a transient rise in cAMP levels in spores (Bagga and Straney, 2000). This strong correlation of flavonoid signaling the cAMP pathway is in contrast with the use of high nutrient (glucose or amino acids) levels to induce germination. The nutrient-induced germination is not inhibited by the PKA inhibitor (Ruan et al., 1995) and glucose treatment does not significantly increase cAMP levels in spores (Bagga and Straney, 2000). Thus, flavonoid- induced germination appears to utilize the cAMP pathway while nutrient-induced germination uses another signaling pathway to trigger development. Since cAMP pathways have been previously implicated as one of several signal pathways in the germination of fungal spores (d'Enfert, 1997), as well as yeast ascospores (Herman and Rine, 1997), the novelty of the *N. haematococca* system is how host flavonoid recognition has been linked to this common signal pathway. The inability of *Fusarium* species which are not pathogens of legumes to germinate in response to flavonoids (Ruan et al., 1995), underscores the pea pathogen's linkage of flavonoid recognition to the cAMP response as a potential adaptation to its host. A key element in this linkage would be the fungal cellular component which binds the specific flavonoid and modulates the cAMP pathway. Such role is often performed by a trans-membrane receptor which triggers a G-protein mediated stimulation of adenylate cyclase activity. Although a flavonoid receptor protein may be present in *N. haematococca,* experiments have implicated another target for flavonoid modulation of the cAMP pathway. The enzymatic activity of cAMP phosphodiesterase isolated from *N. haematococca* spores displayed a strong inhibition by certain flavonoids (Bagga and Straney, 2000). Inhibition of this enzyme within the spore has the potential to increase cAMP levels by decreasing the turnover of cAMP into AMP. Thus, the flavonoids may induce germination by inhibiting the cAMP phosphodiesterase and thus inducing cAMP-dependent germination. Comparison of the flavonoid specificities for germination and phosphodiesterase inhibition support this model (Bagga and Straney, 2000). Strong inducers of germination generally showed stronger inhibition of phosphodiesterase. For example, the strongest inducer of germination, naringenin, displayed a K_i of 33 μM, comparable the phosphodiesterase K_m for cAMP of 24 μM. Weak germination inducers (flavonols, isoflavones and mono- or di-hydroxy flavones and flavanones) were weak inhibitors of cAMP phosphodiesterase activity. Thus, one target of flavonoid modulation of the cAMP pathway may be inhibition of cAMP phosphodiesterase. It is interesting that the specificity of flavonoid inhibition of this fungal cAMP phosphodiesterase differs from that reported for a mammalian cAMP phosphodiesterase (Bagga and Straney, 2000), suggesting that modification of the enzyme's sensitivity to flavonoids allowed adaptation to those specific flavonoids exuded by the host plant. A significant exception to the model is that pisatin, a strong inducer of germination, was not a strong inhibitor of enzyme activity. Perhaps pisatin interacts with another component of the pathway to stimulate cAMP production. It is possible that the receptor which acts to induce the pisatin-inducible genes also plays a role in stimulating germination. Cloning components in the cAMP pathway will allow the direct testing of these models by gene disruption or replacement through DNA transformation.

4. CONCLUSION

The coevolution of pathogenic fungi and plants has produced a narrow range of host specificity for most fungi. The adaptation of the pathogen to its host involves not only the production of host-specific toxins and antagonistic traits, but also in the ability to recognize the presence of the host plant. Such host recognition appears to be more prevalent in soil-borne pathogens than ones which infect aerial parts of the plant through airborne spores. For example, some host recognition is provided by germination and developmental triggers in the cutin covering leaves and stems (Hedge and Kolattukudy, 1997). A greater degree of host recognition in soil-borne fungi may be due to its infection cycle where propagules remain quiescent until the appropriate host appears. However, the flavonoids prevalent in the epidermis of aerial parts may provide yet uncharacterized signals for pathogens interact with the plant above ground. *N. haematococca* provides a useful model for studying the soil-borne pathogens since many of the others lack a sexual stages needed for genetic analysis or have not yet been characterized in terms of pathogenicity genes. Further, *N. haematococca* and other closely related *Fusarium* specialization on legumes takes advantage of the well characterized exudation of flavonoids which may provide host-selective chemical cues. Here we have described two flavonoid responses in *N. haematococca* which may mediate host recognition. One is the control of genes required for pathogenicity on pea by pisatin, the pterocarpan phytoalexin of pea. Using the *PDA1* gene for a model, a pisatin-responsive promoter element has been identified, providing a means to clone components which recognize pisatin and regulate transcription In a pisatin-responsive signal pathway. A second, developmental response is the induction of spore germination by *nod*-inducing flavones and flavanones, as well as pisatin. In both cases, the identification of components in the flavonoid signaling pathway has provided a means to modulate the response and test their biological relevance.

5. ACKNOWLEDGMENTS

This research was supported by a grant from the National Science Foundation (MCB-9419266).

6. REFERENCES

Bagga, S., and Straney, D. C., 2000, Modulation of cAMP and phosphodiesterase activity by flavonoids which induce spore germination of *Nectria haematococca*, *Physiol. Molec. Plant Pathol.* **56**:51-61.
Bohm, B. A., 1998, *Introduction to Flavonoids*, Harwood Academic, Australia, pp. 339-364

Ciuffetti, L. M., and VanEtten, H. D., 1996, Virulence of a pisatin demethylase-deficient strain of *Nectria haematococca* is increased by transformation with the pisatin demethylase gene, *Mol. Plant-Microbe Interact.* **9**:787-792.

Coley-Smith, J. R., and King, J. E., 1970, Response of resting structures of root infecting fungi to host root exudates: An example of specificity, In *Root Diseases and Soil-borne Pathogens*, Toussoun, T. A., Bega, R. V., and Nelson, P. E., eds., Univ. of Calif Press, Berkeley, CA, pp. 130-133.

Connolly, M. S., Williams, N., Heckman, C. A., and Morris, P. F., 1999, Soybean isoflavones trigger a calcium influx in *Phytophthora sojae, Fungal Genet. Biol.*, **28**:6-11.

Covert, S. F., Enkerli, J., Mao, V. P., and VanEtten, H. D., 1996, A gene for maakiain detoxification from a dispensable chromosome of *Nectria haematococca, Mol. Gen Genet.* **251**:397-406.

d'Enfert, C., 1997, Fungal spore germination: Insights from the molecular genetics of *Aspergillus nidulans* and *Neurospora crassa, Fungal Genet. Mol. Biol.* **21**:163-172.

Dakora, F. D., Joseph, C. M., and Phillips, D. A., 1993, Alfalfa (*Medicago sativa* L.) root exudates contain isoflavonoids in the presence of *Rhizobium meliloti, Plant Physiol.* **101**:819-824.

Del Sorbo, G., Andrade, A., Nistelrooy, J., van Kan, J., van Balzi, E., and de Waard, M., 1997, Multidrug resistance in *Aspergillus nidulans* involves novel ATP-binding cassette transporters, *Molec. Gen. Genet.* **254**:4.

Denny, T. P., Matthews, P. S., and VanEtten, H. D., 1987, A possible mechanism of nondegradative tolerance of pisatin in *Nectria haematococca* MP VI, *Physiol. Mol. Plant Pathol.* **30**:93-107.

Dharmatilake, A. J., and Bauer, W. D., 1992, Chemotaxis of *Rhizobium meliloti* towards nodulation gene inducing compounds from alfalfa roots, *Appl. Environ. Microbiol.* **58**:1153-1158.

Firmin, U., Wilson, K. E., Rossen, L., and Johnston, A. W. B., 1986, Flavonoid activation of nodulation genes in *Rhizobium* reversed by other compounds present in plants, *Nature* 324:90-92.

George, H. L., 1993, *Biochemical Characterization of Pisatin Demethylases from Fungal Pea Pathogens*, Ph.D. Thesis. Cornell Univ.

George, H. L., Hirschi, K. D., and VanEtten, H. D., 1998, Biochemical properties of the products of cytochrome P450 genes (PDA) encoding pisatin demethylase activity in *Nectria haematococca, Arch. Microbiol.* **170**:147- 154.

Gheysen, G., 1998, Chemical signals in the plant-nematode interaction, In *Phytochemical Signals and Plant-Microbe Interactions*, Romeo, J. T., Downum, K. R., and Verpoorte, R., eds., Plenum Press, NY., pp. 95-117

Gianinazzi-Pearson, V., Branzanti, B., and Gianinazzi, S., 1989, *In vitro* enhancement of spore germination and early hyphal growth of a vesicular-arbuscular mycorrhizal fungus by host root exudates and plant flavonoids, *Symbiosis* 7:243-255.

Gonzalez-Candelas, L., and Kolattukudy, P. E., 1992, Isolation and analysis of a novel inducible pectate lyase gene from the phytopathogenic fungus *Fusarium solani* f. sp. *pisi* (*Nectria haematococca*, mating population VI), *J. Bact.* **174**:6343-6349.

Graham, T. L., 1991, Flavonoid and isoflavonoid distribution in developing soybean seedling tissues and in seed and root exudates, *Plant Physiol.* **95**:594-603.

Guo, W., Gonzalez-Candelas, L., and Kolattukudy, P. E., 1995, Cloning of a new pectate lyase gene *pelC* from *Fusarium solani* f. sp. *pisi* (*Nectria haematococca*, mating type VI) and characterization of the gene product expressed in *Pichia pastoris, Arch Biochem Biophys.* **323**:352-360

Harborne, J. B., 1971, Distribution of the flavonoids within the *Leguminosae*, in *Chemotaxonomy of the Leguminosae*, Harborne, J. B.,. Boulter, D., and Turner, B. L., eds., Academic Press, NY, pp. 31-69

Harborne, J. B., and Grayer, R. J., 1993, Flavonoids and insects. In *The Flavonoids: Advances in Research since 1986*, Harborne, J. B., ed., Chapman and Hall, London, pp. 589-618.

Hartwig, U. A., Joseph, C. M., and Phillips, D. A., 1991, Flavonoids released naturally from alfalfa seeds enhance growth rate of *Rhizobium meliloti, Plant Physiol.* **95**:797-803.

Hedge, Y., and Kolattukudy, P. E., 1997, Cuticular waxes relieve self-inhibiton of germination and appressorium formation by the conidia of *Magnaporthe grisea, Physiol. Mol. Plant Pathol.* **51**:75-84.

Herman, P. K., and Rine, J., 1997, Yeast spore germination: A requirement for *Ras* protein activity during reentry into the cell cycle, *EMBO J.* **16**:6171-6181.

Hirschi, K., and VanEtten, H., 1996, Expression of the pisatin detoxifying genes (*PDA*) of *Nectria haematococca in vitro* and *in planta, Mol. Plant-Micobe Interact.* **9**:483-491.

Hungria, M., Johnston, A. W. B., and Phillips, D. A., 1992, Effects of flavonoids released naturally from bean (*Phaseolus vulgaris*) on *nodD*-regulated gene transcription in *Rhizobium leguminosarum* bv. *phaseoli, Mol. Plant-Microb. Interact.* **5**:199-203.

Khan, R., and Straney, D. C., 1999, Regulatory signals influencing expression of the PDA1 gene of *Nectria haematococca* MPVI in culture and during pathogenesis of pea, *Molec. Plant- Microbe Interact.* **12**:733-742.

Kistler, H. C., Meinhardt, L. W., and Benny, U., 1996, Mutants of *Nectria haematococca* created by site directed chromosome breakage are greatly reduced in virulence, *Mol. Plant-Micobe Interact.* **9**:804-809.

Krishnan, H. B., and Pueppke, S. G., 1993, Flavonoid inducers of nodulation genes stimulate *Rhizobium fredii* USDA257 to export proteins into the environment, *Mol. Plant-Microbe Interact.* **6**:107-113.

Lane, G. A., Sutherland, O. R. W., and Skipp, R. A., 1987, Isoflavonoids as insect feeding deterrents and antifungal components from root of *Lupinus angustifolius*, *J. Chem. Ecol.* **13**:771-783.

Larson, T. G., and Nuss, D. L., 1994, Altered transcriptional response to nutrient availability in hypovirus-infected chestnut blight fungus, *EMBO J.* **13**:5616-5623.

Li, D., Chung, K. R., Smith, D. A., and Schardl, C. L., 1995, The *Fusarium solani* gene encoding kievitone hydratase, a secreted enzyme that catalyzes detoxification of a bean phytoalexin, *Molec. Plant-Microbe Interact.* **8**:388-397

Lin, T. S ., and Kolattukudy, P. E., 1978, Induction of a biopolyester hydrolase (cutinase) by low levels of cutin monomers in *Fusarium solani* f. sp. *pisi*, *J. Bact.* **133**:942-951.

Lynn, D. G., and Chang, M., 1990, Phenolic signals in cohabitation: Implications for plant development, *Ann. Rev. Plant Physiol.* **41**:497-526.

Maloney, A. P., and VanEtten, H. D., 1994, A gene from the fungal plant pathogen *Nectria haematococca* that encodes the phytoalexin detoxifying enzyme pisatin demethylase defines a new cytochrome P450 family, *Mol. Gen. Genet.* **243**:506-514.

Miao, V. P., Covert, S. F., and VanEtten, H. D., 1991, A fungal gene for antibiotic resistance on a dispensable ("B") chromosome, *Science* **254**:1773-1776.

Mo, Y. Y., Geibel, M., Bonsall, R. F., and Gross, D. C., 1995, Analysis of sweet cherry (*Prunus avium* L.) leaves for plant signal molecules that activate they *SyrB* gene required for synthesis of phytotoxin, syringomycin, *Plant Physiol.* **107**:603-612.

Morris, P. F., and Ward, E. W. B., 1992, Chemoattraction of zoospores of the soybean pathogen, *Phytophthora sojae* by isoflavones, *Physiol. Mol. Plant Pathol.* **40**:17-22.

Morrissey, J. P., and Osbourne, A. E., 1999, Fungal resistance to plant antibiotics as a mechanism of pathogenesis, *Microbiol. Molec. Biol. Rev.* **63**:708-724.

Nelson, E. B., 1991, Exudate molecules initiating fungal responses to seeds and roots, In *The Rhizosphere and Plant Growth*, Keister, L., and Cregan, P. B., eds., Kluwer, Amsterdam., pp. 197-209

O'Donnell, K., and Gray, L. E., 1995, Phylogenetic relationships of the soybean sudden death syndrome pathogen *Fusarium solani* f. sp. *phaseoli* inferred from rDNA sequence data and PCR primers for its identification. *Mol. Plant-Microbe Interact.* **8**:709-716.

Oeser, B., and Yoder, O. C., 1994, Pathogenesis by *Cochliobolus heterostrophus* transformants expressing a cutinase-encoding gene from *Nectria haematococca*, *Mol. Plant Microb. Interact.* **7**:282-289.

Phillips, D. A., Joseph, C. M., and Maxwell, C. A., 1992, Trigonelline and stachydrine released from alfalfa seeds activate *NodD2* protein in *Rhizobium meliloti*, *Plant Physiol.* **99**:1526-1531.

Phillips, D. A., and Kapulnik, Y., 1995, Plant isoflavonoids, pathogens and symbionts, *Trends Microbiol.* **3**:58- 63.

Pieterse, CM. J., Derksen, A.-M. C. E., Folders, J., and Govers, F., 1994, Expression of the *Phytophthora infestans ipiB* and *ipiO* genes *in planta* and *in vitro*, *Mol. Gen. Genet.* **244**:269-277.

Poulin, M. J., Simard, J., Catford, J. G., Labrie, F., and Piché, Y., 1997, Response of symbiotic endomycorrhyza fungi to estrogens and antiestrogens, *Mol. Plant-Micobe Interact.* **10**:481-487.

Pueppke, S. G., 1996, The genetic and biochemical basis for nodulation of legumes by rhizobia, *Crit. Rev. Biotechnol.* **16**:1-5 1.

Ruan, Y., Kotraiah, V., and Straney, D. C., 1995, Flavonoids stimulate spore germination in *Fusarium solani* pathogenic on legumes in a manner sensitive to inhibitors of cAMP-dependent protein kinase, *Mol. Plant-Microbe Interact.* **8**:929-938.

Ruan, Y., and Straney, D. C., 1996, Identification of elements in the *PDA1* promoter of *Nectria haematococca* necessary for a high level of transcription *in vitro*, *Mol. Gen. Genet.* **250**:29-38.

Schäfer, W., Straney, D., Ciuffetti, L., VanEtten, H. D., and Yoder, O. C., 1989, One enzyme makes a fungal pathogen, but not a saprophyte, virulent on a new host plant, *Science* **246**:247-249.

Schmidt, P. E., Parniske, M., and Werner, D., 1992, Production of the phytoalexin glyceollin I by soybean roots in response to symbiotic and pathogenic infection, *Bot. Acta* **105**:18-25.

Straney, D. C., Ruan, Y., and He, J., 1994, *In vitro* transcription and binding analysis of promoter regulation by a host-specific signal in a phytopathogenic fungus, *Anton. van Leeuw.* **65**:183-189.

Straney, D. C., and VanEtten, H. D., 1994, Characterization of the *PDA1* promoter of *Nectria haematococca* and identification of a region that binds a pisatin-responsive DNA binding factor. *Mol. Plant-Microb. Interact.* **7**:256-266.

Talbot, N. J., Ebbole, D. J., and Hamer, J. E., 1993, Identification and characterization of *MPG1*, a gene involved in pathogenicity from the rice blast fungus *Magnaporthe grisea*, *Plant Cell* **5**:1575-1590.

Talbot, N. J., McCafferty, H. R. K., Ma, M., Moore, K., and Hamer, J. E., 1997, Nitrogen starvation of the rice blast fungus *Magnaporthe grisea* may act as an environmental cue for disease expression development, *Physiol. Mol. Plant Pathol.* **50**:179-195.

Tsai, S. M., and Phillips, D. A., 1991, Flavonoids released naturally from alfalfa promote development of symbiotic *Glomus* spores *in vitro*, *Appl. Environm. Microbiol.* **57**:1485-1488.

Tyler, B. M., Wu, M. H., Wang, J. M., Cheung, W., and Morris, P. F., 1996, Chemotactic preferences and strain variation in the response of *Phytophthora sojae* zoospores to host isoflavones, *Appl. Environ. Microbiol.* **62**:2811-2817.

van den Ackerveken, G. F. J. M., Dunn, R. M., Cozijnsen, T. J., Vossen, P., Van den Broek, H. W. J., and de Wit, P. J. G. M., 1994, Nitrogen limitation induces expression of the avirulence gene *avr9* in the tomato pathogen *Cladosporium fulvum*, *Mol. Gen. Genet.* **243**:277-295.

van Gijsegem, F., 1997, *In planta* regulation of phytopathogenic bacteria virulence genes: relevance of plant-derived signals, *European J. Plant Pathol.* **103**:291-301.

VanEtten, H., Funnell-Baerg, D., Wasmann, C., and McCluskey, K., 1994, Location of pathogenicity genes on dispensable chromosomes in *Nectria haematococca* MPVI, *Antonie Van Leeuwenhoek* **65**:263-267.

VanEtten, H. D., 1973, Differential sensitivity of fungi to pisatin and phaseollin, *Phytopathology* **63**:1477- 1482.

VanEtten, H. D., and Kistler, H. C., 1988, *Nectria haematococca* mating populations I and VI, *Adv. Plant Pathol.* **6**:189-206.

VanEtten, H. D., Matthews, D. E., and Matthews, P. S., 1989, Phytoalexin detoxification: Importance for pathogenicity and practical considerations, *Ann. Rev. Phytopathol.* **27**:143-164.

Wasmann, CC., and VanEtten, H. D., 1996, Transformation-mediated chromosome loss and disruption of a gene for pisatin demethylase decrease the virulence of *Nectria haematococca* towards pea, *Mol. Plant-Microbe Interact.* **9**:793-803.

Wengelnik, K., Marie, C., Russel, M., and Bonas, U., 1996, Expression and localization of *HrpA1*, a protein of *Xanthomonas campestris* pv. *vesicatoria* essential for pathogenicity and induction of the hypersensitive reaction, *J. Bact.* **178**:1061-1069.

Winans, S. C., 1992, Two way signaling in *Agrobacterium*-plant interactions, *Microbiol. Rev.* **56**:12-31.

Xiao, Y., Lu, Y., and Hutcheson, S. W., 1992, Organization and environmental regulation of the *Pseudomonas syringae* pv *syringae 61 hrp* cluster, *J. Bact.* **174**:1734-1741.

Yokosawa, R., Kunninaga, S., and Sekizaki, H., 1986, *Aphanomyces euteiches* zoospore attractant isolated from pea root prunetin, *Ann. Phytopath. Soc. Japan* **52**:809-816.

SIGNALLING IN ARBUSCULAR MYCORRHIZA: FACTS AND HYPOTHESES

Horst Vierheilig[1] and Yves Piché[1]

1. INTRODUCTION

The arbuscular mycorrhizal symbiosis is an association between plant roots and fungi. Arbuscular mycorrhizal fungi (AMF) colonize roots improving plant nutrition mainly by transferring phosphate (P) from the soil to the plant, whereas plants provide the fungi with carbohydrates (Smith and Read, 1997). In contrast to the rhizobial symbiosis with a host range limited to the *Leguminoseae*, AMF form symbiotic associations with a wide range of plant species. Interestingly, there seem to be striking similarities between signalling in rhizobial and arbuscular mycorrhizal symbiosis (reviewed by Hirsch and Kapulnik, 1998). Apart from the effect of plant derived secondary plant compounds (SPC) on the bacterial and the fungal symbiont, SPC (e.g. flavonoids) are accumulated in the roots of the respective host plants during the establishment of both symbioses. Whereas there is some information on the role of SPC in the rhizobial symbiosis, the exact role of SPC during the establishment of the AM symbiosis still remains unclear.

The data about flavonoids in the arbuscular mycorrhizal symbiosis have been extensively reviewed (Vierheilig et al., 1998a). Although there is new information on the effect of various flavonoids on the growth of arbuscular mycorrhizal fungi, and the induction of flavonoids during root colonization by arbuscular mycorrhizal fungi, there is still a debate on the exact role of flavonoids during the formation of the arbuscular mycorrhizal symbiosis. Based on current work, we hypothesize that a fundamental molecular dialogue occurs that regulates not only the early development of AM symbiosis, but also subsequent colonization which has to be balanced for establishment of genuine mycorrhizal symbiosis. In this work some aspects of the signalling and regulation in rhizobial and arbuscular mycorrhizal symbiosis are compared and some possible functions of secondary plant compounds e.g.

---

[1] Centre de Recherche en Biologie Forestière (CRBF), Pavillon C.- E.- Marchand, Université Laval, Québec, G1K 7P4, Canada.

Flavonoids in Cell Function, edited by Béla Buslig and John Manthey.
Kluwer Academic/Plenum Publishers, New York, 2002.

flavonoids, in the mycorrhizal symbiosis are presented. We hope this overview gives some stimulation for further studies on signalling in AM.

2. THE LEGUMES-RHIZOBIA INTERACTION

2.1. Signalling During the Formation of the Rhizobial Symbiosis

Abundant information is available about signalling in the legume-rhizobium interaction. Root exudates of legumes seem to stand at the beginning of a complex signal exchange cascade (Fig. 1). Root exudates contain a wide range of compounds, however, during the formation of the rhizobial symbiosis secondary plant compounds, specifically flavonoids (reviewed by Phillips and Tsai, 1992), play a key role. Flavonoids act as chemoattractants for the rhizobial bacteria and as specific inducers of rhizobial nodulation genes (*nod*-genes), which are involved in the synthesis of lipo-chitooligosaccharide signals, called *Nod*-factors (recently reviewed by Perret et al., 2000). *Nod*-factors induce in roots of certain legumes the accumulation of flavonoids resulting in the secretion of more flavonoids by the root, which in turn stimulate further production of *Nod*-factors by the bacteria (Recourt et al., 1992; Dakora et al., 1993; Schmidt et al., 1994; Bolanos-Vasquez and Werner, 1997). *Nod*-factors are required for the rhizobial penetration of roots hairs of the host plant. After penetration of the root, the bacterium reaches through an infection thread the root cortex, where dividing cells form a new symbiotic organ, the nitrogen-fixing nodule.

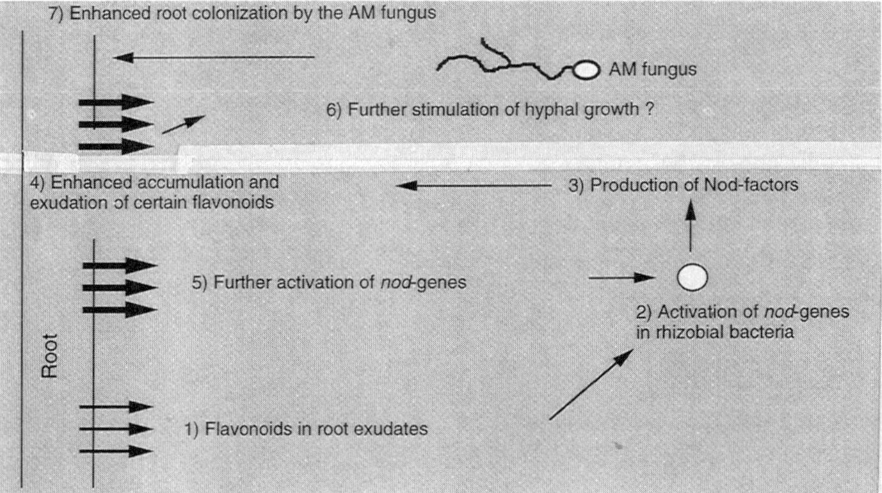

Figure 1. Schematic model of the signalling in the legume-*Rhizobium*-arbuscular mycorrhizal fungus (AMF) interaction. The legume-*Rhizobium* interaction (1-4). The legume-*Rhizobium*-AMF interaction (6-7). Flavonoids are exuded by roots of legumes (1) activating *nod*-genes in the rhizobial bacteria (2). The bacteria start to produce *Nod*-factors (3) which are perceived by the legume root and induce the accumulation and exudation of flavonoids by the root (4). The flavonoids again activate *nod*-genes (5). The changes in the root exudation also seem to enhance root colonization by AMF (7), possibly through a stimulatory effect on hyphal growth of AMF (6).

2.2. Autoregulation of Nodulation

Ample evidence indicates that the rhizobial host plant can control the extent of nodulation (for details see review Caetano-Anollés and Gresshoff, 1991). Once a critical number of nodules has formed on the root system further nodulation is nearly completely suppressed. The formation of nodules is a costly process for the host, so it is not surprising to find that nodulation is controlled by a plant-mediated autoregulatory mechanism. Interestingly, the autoregulatory mechanism is systemic. Inoculation of one half of a split root system with *rhizobia* strongly reduces nodulation on the other half of the root system, whereas the total number of nodules formed on a host plant remains constant. Little is known about the mechanisms and the signals involved in the local and systemic auto-regulation of nodulation.

2.3. The Legume-*Rhizobia*-AMF Interaction

In many studies an enhanced root colonization by AMF and rhizobia has been reported when co-inoculated (Daft and El-Giahmi, 1974; Cluett and Boucher, 1983; Kawai and Yamamoto, 1986; Pacovsky et al., 1986; Chaturvedi and Singh, 1989; Xie et al., 1995). A more detailed study with soybean mutants which are unable to form nodules (*Nod*) revealed that root colonization by AMF was even increased when *Nod* soybeans were inoculated with rhizobia (Xie et al., 1995). These data clearly demonstrated that the stimulatory effect on AM root colonization by rhizobia is not linked to nitrogen fixation but rather to a pre-nodulation event. When soybean plants were co-cultivated with AMF and a rhizobial mutant strain, which was deficient in *Nod*-factor biosynthesis and thus non-nodulating (*Nod*) or a rhizobial strain (*Nod*⁺) with a normal *Nod*-factor biosynthesis, the rhizobial *Nod*⁺ strain showed a clearly stimulatory effect on AMF colonization, whereas the rhizobial *Nod* strain showed no effect. These results suggested a stimulatory effect of *Nod*-factors on AM root colonization. However, testing two *Nod*-factors differing in their ability to induce the secretion of flavonoids in soybean, the *Nod*-factors could be excluded as the responsible factors for the increased root colonization. Only the *Nod*-factor enhancing the secretion of flavonoids from roots showed a stimulatory effect on AMF root colonization. The *Nod*-factor having no effect on flavonoid secretion did not affect AM root colonization (Xie et al., 1995). A similarly enhanced AM colonization was observed when *Nod*-factors were applied to *Lablab purpureus,* another leguminous species (Xie et al., 1998). These results suggest that certain flavonoids induced in the signal exchange cascade during rhizobial symbiosis formation do have a stimulatory effect on the establishment of the AM symbiosis (see Fig. 1). The application of several *Nod*-factor-induced flavonoids from soybean to AMF inoculated soybean plants also enhanced AM root colonization (Xie et al., 1995), which seems to link the enhanced AMF colonization in presence of rhizobia to an enhanced root exudation of certain flavonoids. The stimulatory effect of flavonoids on root colonization has been shown in a variety of plants (reviewed Vierheilig et al., 1998a; Fig. 2).

To our knowledge no data exist to show whether these changes in root exudation in rhizobial roots also affect other microorganisms in the soil. Legumes inoculated with rhizobia are less damaged by soil borne pathogens (Chakraborty and Purkayastha, 1984; Chakraborty and Chakraborty, 1989; Tu, 1978). No evidence has been presented so far to explain this protective effect conclusively. However, as flavonoids also exhibit antifungal

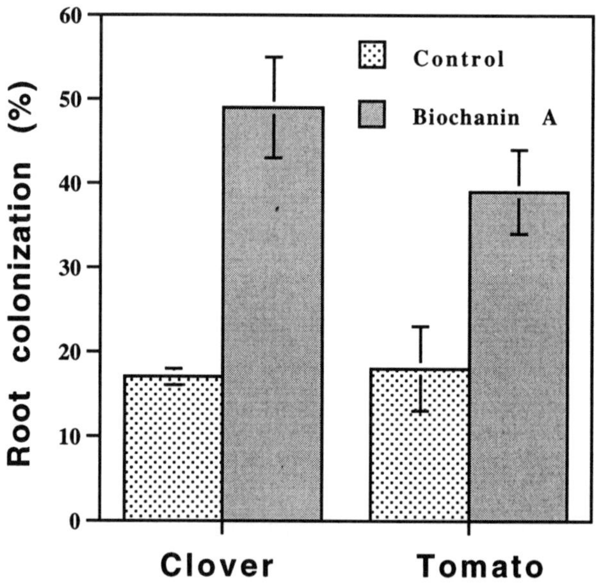

Figure 2. Effect of Biochanin A on root colonization of clover and tomato by the AMF, *Glomus mosseae*. Plantlets (10 d old) were transferred into pots containing an autoclaved sand/loam substrate (1:1; v:v) mixed with an inoculum of *G. mosseae* (containing sporocarps, spores, hyphae and colonized root pieces). Biochanin A (disolved in MeOH; final concentration 0.5%) was applied at time of transplanting and two weeks later. After four weeks roots were harvested and root colonization was determined.

activites (Van Etten, 1976; Wyman and Van Etten, 1978; Weidenbörner and Jha, 1994; Dakora and Phillips, 1996), changed concentrations of these compounds in root exudates of rhizobial plants could affect soil borne plant pathogens.

These results indicate that changes in the root exudates occur, which not only play a key role during the formation of the rhizobial symbiosis, but which also affect AMF and possibly other soil microorganisms.

3. THE PLANT-AMF INTERACTION

3.1. The Precolonization Stage (Signals from the Plant to the Fungus)

The association between legumes and rhizobia is perhaps the most extensively studied plant-microbe interaction, and the signalling between the host and bacterium is fairly well understood. Little is known, however, about signalling during the formation of the AM symbiosis, although numerous publications exist. Data on the effect of various compounds on hyphal growth and on the accumulation of compounds in mycorrhizal roots are available (recently reviewed by Vierheilig et al., 1998a), however, the significance of these events for the outcome of the symbiosis remains unclear. This lack of understanding is partially due to the difficulty to work with AMF. In contrast to rhizobia, AMF can not be cultured in the

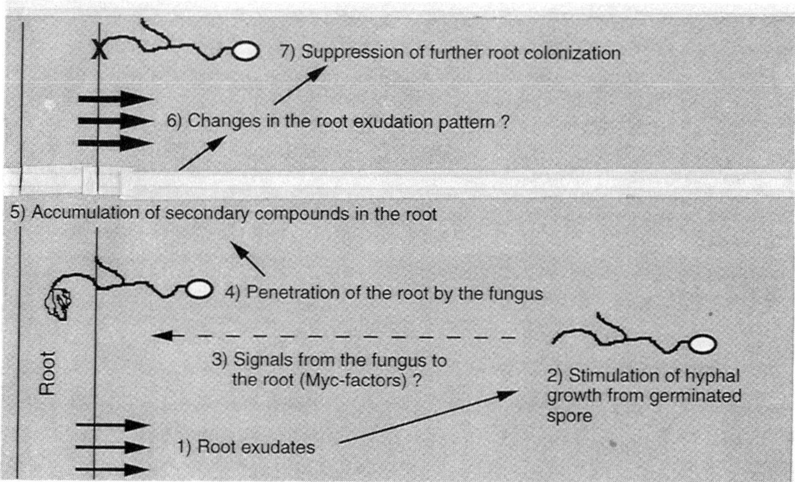

Figure 3. Schematic model of the signalling in the plant-AMF-AMF interaction. The plant-AMF interaction in a non-mycorrhizal root (1-5). The plant-AMF-AMF interaction in a mycorrhizal root (6-7). Root exudates (1) stimulate hyphal growth of AMF (2). Possibly fungus-derived signals are perceived by the plant, before root colonization (3). The fungus penetrates the root forming its intraradical structures (4) and induces the accumulation of secondary plant compounds (SPC) in the mycorrhizal root (5). Changes in the mycorrhizal root occur *e.g.*, the root exudation is altered (6), possibly affecting further root colonization negatively (7).

absence of a host plant, thus it can be difficult to distinguish between plant and fungal effects. Most data are available about signals originating from the plant during the precolonization stage, i. e., the phase before root colonization occurs.

Similar to the legume-rhizobia interaction, root exudates appear to be at the beginning of a signal exchange chain between the AM host plant and the AMF (Fig. 3). Root exudates exhibit an attractional effect on the growth of AMF (Gemma and Koske, 1988; Koske, 1982; Koske and Gemma, 1992; Suriyapperuma and Koske, 1995; Vierheilig et al., 1998b). The nature of the attracting compounds is still unidentified. More information is available about the compounds in root exudates affecting AMF growth and spore germination (see review Vierheilig et al., 1998a). In general, under axenic conditions root exudates of AM host plants exhibited a clearly stimulatory effect. In light of the importance of flavonoids exuded by roots during establishment of rhizobial symbiosis, the effects of these compounds on AMF was tested early. In 1989 (Gianinazzi-Pearson et al.) the first report was published on AM hyphal growth stimulation by the flavonoids hesperetin, apigenin and naringenin. A short time later a synergistic stimulatory effect of flavonoids and enhanced CO_2 concentration on AM hyphal growth was reported (Bécard et al., 1992; Chabot et al., 1992; Poulin et al., 1993). Thereafter many flavonoids and other SPC were tested for their effects on axenic growth of AMF, some exhibiting a clearly stimulatory effect and some exhibiting an inhibitory or no effect a all (see Vierheilig et al., 1998a).

The effect of root exudates and flavonoids was also tested on root colonization by AMF. Interestingly application of root exudates of non-mycorrhizal plants (Tawaraya et al., 1998) or of certain flavonoids to AMF inoculated plants enhanced root colonization (see review by Vierheilig et al., 1998a; Fig. 2).

Although the data on the stimulatory effect of root exudates and flavonoids on axenic growth and root colonization by AMF are certainly interesting, their biological relevance to the outcome of the mycorrhizal symbiosis still has to be demonstrated. Bécard et al. (1995) recently questioned an essential role of flavonoids during the establisment of the AM association.

3.2. The Precolonization Stage (Signals from the Fungus to the Plant)

To our knowledge, due to the inability to culture AMF axenically, no direct evidence of the presence of fungal derived signals has been reported as yet. However, there is some support for the view that AMF release signalling compounds which can be perceived by the root and which are probably essential for root colonization by AMF.

In general, plants in the *Brassicaceae* familiy are nonhosts for AMF (Harley and Harley, 1987). The inability of these plants to form the symbiosis is characterized by the absence of fungal structures in the root or even on the root surface. In roots of *Brassicacea* inoculated with AMF changes of the ß-1,3-glucanase and chitinase activity (Vierheilig et al., 1994), and changes of the glucosinolate levels could be observed (Vierheilig et al., 2000a). The biological function of these changes is not clear, however, these observations point to the presence of AM fungal derived signals sensed by plants. The non-specific character of these factors is suggested since even AM nonhost plants seem to sense the presence of AMF. A so-called "*Myc*-factor" (analogous to the *Nod*-factor known from rhizobia) recently has been proposed by Albrecht et al. (1998) and Blilou et al. (1999). Albrecht et al. (1998) found a similar induction of early noduline genes in pea roots by an AMF and by rhizobial *Nod*-factors, and Blilou et al. (1999) reported a similar suppression of salicylic acid in pea by *Nod*-factors and AMF.

Xie et al. (1999) detected a specific chitinase isoform in soybean roots, which was only induced in mycorrhizal roots and in roots treated with *Nod*-factors, but not in non-mycorrhizal untreated roots. Interestingly, a chitinase isoform found in mycorrhizal clover roots was also induced under sterile conditions when AM mycelium was applied to clover roots (C. Albrecht and H. Vierheilig, unpublished results), pointing to a certain functional similarity of signals in the fungal mycelium with *Nod*-factors. This would not be surprising as *Nod*-factors share structural characteristics with fungal chitin elicitors. However, direct evidence for the presence of "*Myc*-factor" is still missing.

3.3. Penetration and Root Colonization

After the AMF has reached the root, hyphal growth on the root surface and the formation of penetration structures, the appressoria, occur. Finally the fungus penetrates the root, forming internal hyphae and arbuscules. The molecular mechanisms controlling the penetration, the establishment and the functioning of the symbiosis are still unclear. SPC are thought to play a role during these events, as the accumulation of several SPC has been reported in mycorrhizal plants (see review Vierheilig et al., 1998a).

The most extensive information on the accumulation pattern of SPC in mycorrhizal plants is available about isoprenoid cyclohexenone derivatives (CD). Several studies of roots of many members of the *Poaceae* colonized by AMF showed the widespread occurrence of glycosylated C13 cyclohexenone derivatives in this plant family (Maier et al., 1995, 1997; Peipp et al., 1997; Vierheilig et al., 2000b). Recently CD have been detected in mycorrhizal roots of *Nicotiana tabacum* (Maier et al., 1999, 2000) and *Lycopersicon esculentum* (Maier et al., 2000), both members of the *Solanaceae*, showing that their occurrence is not restricted to the *Poaceae*. CD accumulation is specifically induced by AMF, but not by root pathogens or other endophytes (Maier et al., 1997), and is directly correlated with the degree of mycorrhization (Maier et al., 1995). Various AMF from different species or genera induce qualitatively similar CD accumulation patterns (Vierheilig et al., 2000b). Although there is some information on the accumulation of CD in mycorrhizal roots, nothing is known yet about the function of these compounds in mycorrhiza.

Fewer studies exist about the accumulation of flavonoids, another group of SPC, in mycorrhizal roots. However, abundant data are published about their effect, and thus their possible function, on AMF (Vierheilig et al., 1998).

A number of flavonoids could be identified in mycorrhizal roots (Harrison et al., 1993; Volpin et al., 1994). In fully colonized *Medicago sativa* plants, increased levels of 4',7-dihydroxyflavone, formononetin, coumestrol and daidzein could be detected (Harrison et al., 1993). In roots, where root colonization was halted at the appressoria formation stage 4',7-dihydroxyflavone and coumestrol was not detectable, but formononetin and daidzein accumulated. These observations pointed to accumulation of certain compounds at different stages of the root colonization process and could indicate their function at different steps during the formation of the symbiosis.

The similarity between the rhizobial and the mycorrhizal symbiosis is striking. In both symbioses SPC are accumulated in the roots. However, where in rhizobial symbiosis flavonoids accumulated are released by the root interacting with the bacterium, the probable function of flavonoids and CD in mycorrhizal roots is speculative. It is interesting to hypothesize how secretion of these SPC by the mycorrhizal root affects further mycorrhization.

4. THE PLANT-AMF-AMF INTERACTION

4.1. Is There an Autoregulation of Mycorrhization?

The AM association is not a static event. After a first colonization of a root by an AMF, the same AMF or other AMF can colonize the same root or other parts of the root system of the same plant. From the plant's perspective, the development of the mycorrhizal association is a costly process Thus, after a critical degree of root colonization is reached, suppression of further root colonization by AMF is likely to occur in order to reduce the cost of symbiosis for the plant as a carbon source. Such an autoregulatory mechanism is known in rhizobial symbiosis (see 2.2 and Caetano-Anollés and Gresshoff, 1991).

Results on this hypothesis have been recently published. In several studies simultaneous inoculation with different AMF reduced root colonization of each individual AMF (Daft and Hogarth, 1983; Wilson, 1984; Hepper et al., 1988; Pearson et al., 1993, 1994). However, under these conditions a faster root colonizer could simply dominate a slower root colonizer by occupying the biological niche in the root, thus reducing root colonization by the slow colonizer. To address this possibility, the effect of precolonization by an AMF on further root colonization by AMF was tested in a recent experiment. One side of a split root system of barley plants was inoculated with *G. mosseae*, *G. intraradices* or *Gigaspora rosea*. When the fungi extensively colonized the roots, the other side of the split root system was inoculated with *G. mosseae*. The results showed a clear reduction of subsequent root colonization once the barley plants were precolonized with any of the tested AMF (Vierheilig et al., 2000c), indicating a systemic suppression of further root colonization of plants colonized by an AMF. The effect was also observed when *G. mosseae* precolonized plants were subsequently inoculated with *G. mosseae*, pointing to an autosuppression of further root colonization by the same fungus (Vierheilig et al., 2000c).

Pearson et al. (1993) suggested several mechanisms to explain the reduction of AMF colonization in roots colonized by another AMF: i) a competition of different AMF for carbohydrates, ii) a mechanism involving the improved P-status of AMF colonized plants, iii) an altered accumulation of compounds in mycorrhizal roots affecting AMF.

A comparison of the root growth in mycorrhizal and non-mycorrhizal split-root systems and the observation that subsequent root colonization by *G. mosseae* was suppressed in *G. mosseae* precolonized plants makes it rather unlikely that the observed suppression in mycorrhizal plants was due to a simple competition for carbohydrates between AMF (Vierheilig et al., 2000c). Moreover, P-application experiments in split root systems demonstrated that improved P-status and the suppressional effect are not linked in mycorrhizal plants (Vierheilig et al., 2000d).

Changes in the accumulation pattern of flavonoids resulting in an altered root exudation have been reported from the rhizobial symbiosis in legumes (see 2.1). Moreover, it has been reported that changes in the root exudation pattern during the formation of the rhizobial symbiosis in legumes can affect root colonization by AMF (Xie et al., 1995; 1998). Thus it is reasonable to speculate that root exudates of mycorrhizal plants exhibit a different effect on AMF than root exudates of non-mycorrhizal plants. Pinior et al. (1999) showed that root exudates of cucumber plants colonized by the AMF *G. mosseae*, *G. intraradices* or *Gi. rosea* applied to cucumber plants inoculated by *G. mosseae* exhibited a different effect on root colonization then root exudates of non-mycorrhizal cucumber plants. All root exudates from mycorrhizal plants showed a clearly inhibitory effect on root colonization by *G. mosseae*, and root exudates of non-mycorrhizal cucumber plants slightly stimulated root colonization. The inhibitory effect of mycorrhizal root exudates on root colonization was similar, independent of the root colonizing fungus. Root exudates from plants colonized by *G. mosseae* also inhibited root colonization by *G. mosseae*. A similar autosuppressional effect has been observed in split root systems as discussed above (Vierheilig et al., 2000c).

From the data presented an altered host status of mycorrhizal and non-mycorrhizal roots is suggested. Once the roots are mycorrhized, further root colonization by AMF seems suppressed, indicating an autoregulation of the mycorrhization. Changes in the root exudation pattern of mycorrhizal roots seem to be at least partially involved in this suppression.

4.2. Which Compounds Might Be Involved in the Autoregulation of Mycorrhization?

Several SPC are accumulated after root colonization by AMF. The importance of these SPC for the outcome of the AM symbiosis is still unclear. Possibly, the role of SPC, e.g., flavonoids, changes during the formation of the mycorrhizal association. Initially, SPC exuded by the root stimulate and attract AMF hyphal growth. Subsequently, AMF-derived signals ("*Myc*-factors"?) induce the accumulation and secretion of certain SPC before a cell-to-cell contact and/or at the moment of the appressoria formation, thus stimulating further root colonization. Finally, the fungus penetrates the root, forming its intraradical structures and inducing the accumulation of SPC which are subsequently released into the rhizosphere, thus suppressing further root colonization.

The stimulation of AMF by SPC found in root exudates of non-mycorrhizal plants is fairly well documented (see Vierheilig et al., 1998a). However, there is little evidence to support the other possible functions of SPC proposed in our model, that different flavonoids act and accumulate prior or during different stages of the symbiosis. Although there are indications for the presence of "*Myc*-factors" (see 3.2), no data are available on the accumulation of SPC induced by these fungus derived signals. Harrison and Dixon (1993) found that some flavonoids are accumulated at the cell-to-cell stage (formononetin and daidzein), before the plant-AMF association is established, whereas others are not (4',7-dihydroxyflavone and coumestrol). While 4',7-dihydroxyflavone, formononetin, coumestrol and daidzein are known to stimulate spore germination or hyphal growth of AMF (see review by Vierheilig et al., 1998a), the phytoalexine medicarpin, also accumulated in mycorrhizal *Medicago* roots, possesses antifungal activity toward pathogenic fungi (Higgins, 1978), thus possibly playing a different role in the symbiosis than the other compounds accumulated. Flavonoids, depending on their nature and concentration, can stimulate fungal growth or exhibit antifungal activity (Van Etten, 1976; Wyman and Van Etten, 1978; Weidenbörner and Jha, 1994; Dakora and Phillips, 1996).

As yet, there are no available data that compounds accumulated in mycorrhizal roots are secreted by the root. However, root exudates of mycorrhizal plants suppress root colonization by AMF (Pinior et al., 1999), suggesting the presence of inhibitory compounds. It is interesting to note that different concentrations of phenolic compounds applied to AMF-inoculated plants can exhibit different effects. Low concentrations of some phenolics can stimulate root colonization, while higher concentrations can reduce root colonization (Fries et al., 1997). Thus, it is possible that the accumulation of SPC occurring after root colonization by AMF triggers an enhanced exudation of these compounds in the rhinosphere and affects further root colonization.

The possible role of phenolics accumulated in mycorrhizal roots in the regulation of mycorrhization is also confirmed by results with ectomycorrhizal fungi. Feugey et al. (1999) reported that the accumulation of phenolic compounds after root colonization by an ectomycorrhizal fungus limited further formation of the Hartig net to the outer layer of the root cortex.

Until now a role of phenolics, specifically flavonoids, has been proposed only for the formation of the symbiotic association, however, we suggest that different phenolics or different concentrations of phenolics might play different roles during different stages of the AM symbiosis.

Some studies with SPC involved in the regulation of root colonization were conducted in the *Gramineae*. CD are SPC specifically induced by AMF (Maier et al., 1997), and application of the CD, blumenin, to AMF inoculated plants has been shown to reduce root colonization (Fester et al., 1999).

Vierheilig et al. (2000c) reported that in a split root system precolonization by an AMF on one side suppressed further colonization on the other side. The mechanisms of this suppression are still unknown, however, the accumulation of CD in a mycorrhizal plant might be involved. Studying the accumulation of several CD in split root systems of barley revealed that CD are only accumulated in mycorrhizal roots, but never in non-mycorrhizal roots of a mycorrhizal plant (Vierheilig et al., 2000d). This clearly demonstrated that CD are not systemically induced, and thus can not be involved in the systemic autoregulation of mycorrhization first reported by Vierheilig et al. (2000c). However, a local effect of CD on mycorrhization can not be excluded based on the results presented, although, the lack of antifungal activity of CD on the growth of the fungus *Cladosporium cucumerinum* (Fester et al., 1999) makes a simple growth inhibitory effect on AMF rather unlikely.

The results presented indicate that SPC are involved during different stages of the formation of the AM association. We attempted to consider these results with the possibility that SPC are involved in the regulation of mycorrhization in mycorrhizal plants, however, clear proofs for such a role are still lacking.

4.3. Possible Signalling Compounds in the Autoregulation of Mycorrhization

In systemic acquired resistance (SAR) pre-infection of a plant with a pathogen can result in an enhanced resistance in other parts of the plant to the same or related pathogens (recently reviewed by Sticher et al., 1997). Salicylic (SA) and jasmonic acid (JA), two of the compounds playing a key role in SAR, have been extensively studied (reviewed by Malamy and Klessig, 1992; Pieterse and van Loon, 1999), however, little information is available on either compound in AM (recently reviewed by Ludwig-Müller, 2000).

Transgenic *Nah*G plants have been a valuable tool to study SA mediated plant mechanisms. *Nah*G plants express the bacterial *nahG* gene, which encodes the enzyme salicylate hydroxylase that inactivates SA. Thus SA is not accumulated in *Nah*G plants (Gaffney et al., 1993). Inoculation of *Nah*G tobacco plants with AMF resulted in accelerated root colonization, compared with wild type tobacco plants (J. M. Garrido-Garcia and H. Vierheilig, unpublished results), indicating that SA levels in roots can affect root colonization by AMF. Measurements of the SA accumulation in plants inoculated with AMF also indicated that SA is involved in the susceptibility of plants to AMF. In pea mutants (*Myc*⁻ plants), which are unable to form the AM symbiosis, the SA accumulation was enhanced, whereas in *Myc*⁺ peas the SA accumulation was low (Blilou et al., 1999) or only transient (Blilou et al., 2000a, 2000b) when compared to the *Myc*⁻ pea mutants .

These observations suggest a role of SA in the regulation of root colonization by AMF. However, whether enhanced SA levels are the symptom or the cause of reduced colonization still has to be investigated.

Even less data are available about the SAR signalling compound JA in AM (reviewed by Ludwig-Müller, 2000). Recent experiments show that whereas an application of SA to the shoot of cucumber plants shows no effect on root colonization by AMF, while an application of JA to the shoot drastically reduces root colonization (control root colonization 30±4 %, JA treated root colonization 0.5±0.3%) (J. Ludwig-Müller and H. Vierheilig, unpublished results). Moreover, JA is extensively accumulated in mycorrhizal roots and even in non-mycorrhizal roots of a split-root system of mycorrhizal plants, in general, at levels higher than in roots of non-mycorrhizal plants (O. Miersch and H. Vierheilig, unpublished results)(Table 1). These results may indicate an involvement of JA not only in the local, but also in the systemic suppression of further root colonization by fungi in mycorrhizal plants. From the results available on SA and JA in AM, their importance in the autoregulation of mycorrhi-zation, similar to their role in SAR, can not be excluded. However, their exact function in local and systemic bioprotective and autoregulatory mechanisms in mycorrhizal plants still has to be elucidated.

Table 1. Accumulation of jasmonic acid (JA) in a split root system[a] of non-mycorrhizal (Plants 1 to 3) and of mycorrhizal cucumber plants (Plants 4 to 7). In mycorrhizal plants one half of the split root system was mycorrhizal (+M) and the other half was non-mycorrhizal (-M)

	Treatment	Root colonization (%)	JA (pmol/g FW)	JA (means ± s. e.)
Plant 1	non-mycorrhizal plant	0	314	non-mycorrhizal plants 398 ± 106
Plant 2	non-mycorrhizal plant	0	364	
Plant 3	non-mycorrhizal plant	0	517	
Plant 4	mycorrhizal plant Side -M	0	832	
	mycorrhizal plant Side +M	29	1943	mycorrhizal plants Side -M 627 ± 145
Plant 5	mycorrhizal plant Side -M	0	568	
	mycorrhizal plant Side +M	41	10480	
Plant 6	mycorrhizal plant Side -M	0	496	
	mycorrhizal plant Side +M	64	2178	mycorrhizal plants Side +M 5624 ± 4249
Plant 7	mycorrhizal plant Side -M	0	613	
	mycorrhizal plant Side +M	49	7896	

[a] Five days (d) old cucumber plants were excised of the main root and grown for 5 d. Thereafter plants were transferred into the compartment system described by Vierheilig et al. (2000c) in order to colonize the split root systems by the AM fungus *Glomus mosseae* (BEG 12). Control plants were not inoculated. Sixteen days after the inoculation plants were harvested and the roots were stored at -20°C. The accumulation of JA was determined in lyophilized roots (Miersch and Wasternack 2000). Root colonization was determined as described by Vierheilig et al. (1998c).

5. AUTOREGULATION OF MYCORRHIZATION AND ENHANCED RESISTANCE TOWARD SOIL BORNE FUNGAL PATHOGENS IN MYCORRHIZAL PLANTS: ONE MECHANISM, TWO SYMPTOMS?

In many studies the enhanced resistance of mycorrhizal plants toward soil borne fungal pathogens has been reported (Dehne, 1982). The mechanisms involved in the resistance have not been clearly identified (Azcon-Aguilar and Barea, 1996; Hooker et al., 1994; Morandi, 1996; St.-Arnaud et al., 1995). Although there is no experimental evidence yet for a common mechanism of autoregulation and enhanced resistance, it is tempting to speculate that they share certain features.

An indication for a common mechanism is the effect of root exudates of mycorrhizal plants on pathogens and AMF. The AMF non-host plant *Dianthus caryophyllus* co-cultured with the mycorrhizal AMF host plant *Tagetes patula* reduced disease development by *Fusarium oxysporum* in *D. caryophyllus* (St.-Arnaud et al., 1997) suggesting the presence of bioprotective compounds released by the roots of the mycorrhizal plant. This observation is confirmed by earlier data presented by Caron et al. (1986), Meyer and Linderman (1986) and Bansal and Mukerji (1994), reporting suppression of pathogenic fungi in the soil around mycorrhizal roots.

More detailed studies on root exudates of mycorrhizal and non-mycorrhizal plants revealed some interesting details. In experiments *in vitro*, root exudates of non-mycorrhizal plants exhibited a clearly stimulatory effect on the sporulation of the pathogenic fungus *Phytophthora fragariae* (Norman and Hooker, 2000), and the hyphal growth of AMF (Pinior et al., 1999). However, when root exudates of mycorrhizal plants were collected no stimulation could be observed, pointing to the absence of a stimulatory compound/s or the presence of a inhibitory compound/s in the root exudates of mycorrhizal plants.

Root exudates of mycorrhizal plants seemed not only to affect fungal sporulation and growth, but also reduced the susceptibility of roots to infection by fungi. Working with tomato plants and the soil borne pathogen *P. parasitica*, Vigo et al. (2000) explained a decreased susceptibility of mycorrhizal roots to pathogens by a reduction of infection loci. A similar mechanism could explain the reduced root colonization by AMF in cucumber root, treated with root exudates of mycorrhizal cucumber plants, as these exudates exhibited no direct inhibitory effect on AMF (Pinior et al., 1999).

To summarize, the similar effect of root exudates of mycorrhizal plants on AMF and soil borne pathogens could be explained by changes in the exudates, not only reducing fungal growth, but also reducing possible infection sites on the host roots. Whether the same compounds in the exudates of mycorrhizal roots do affect pathogens and AMF needs further investigation.

Further evidence for a possibly common mechanism is the systemic effect observed in autoregulation and in plant resistance. In several studies a systemic suppression effect of AMF root colonization on soil borne fungal pathogen infection has been demonstrated (Davis and Menge, 1980; Rosendahl, 1985; Cordier et al., 1998). The suppression of further root colonization by AMF in already mycorrhized plants also appears to be regulated systemically (Vierheilig et al., 2000c, 2000d). The signals that control the systemic auto-regulatory mechanism of mycorrhization and the enhanced systemic resistance in mycor-rhizal plants are not yet known. A common signalling mechanism seems possible (see 4).

Another indication for a common mechanism in autoregulation and enhanced resis-tance is that only a well established AM symbiosis can protect plants against soil borne pathogens. In several reports the requirement of extensive root colonization by AMF for en-hanced resistance has been presented (Cordier et al., 1996, 1998; Caron et al., 1986; Bärtschi et al., 1981) and recently it was suggested that a bioprotective effect depends on a fully developed symbiosis characterised by the presence of arbuscules (Slezack et al., 2000). The recently reported systemic autoregulatory mechanisms suppressing further root colonization by AMF in already mycorrhizal plants also seems to depend on the intensity of prior mycorrhization (Vierheilig et al., 2000c, 2000d). High pre-colonization on one side of a split root system results in high reduction of root colonization on the other side, and low pre-colonization results in low reduction of further root colonization (Vierheilig and Piché unpublished results). Thus both phenomena, autoregulation and enhanced resistance, seem to depend on an extensive root colonization by AMF.

Although the presented data can be interpreted differently, it is reasonable to propose "one mechanism, two symptoms" for autoregulation and bioprotection in mycorrhizal plants. It is possible that mycorrhizal plants, while trying to limit the cost of the AM symbiosis also acquired bioprotection against pathogenic fungi. It is plausible that an already mycorrhized plant develops only a single mechanisms to repulse further colonization by fungi, not discriminating between AM fungi and soil borne pathogenic fungi. Although this hypothesis is very exciting, the data available so far can only offer indications for its validity. However, clear proof is still lacking. Further studies are needed to elucidate the exact mechanism for autoregulation of mycorrhization and enhanced resistance toward soil borne pathogens in mycorrhizal plants.

6. CONCLUSION

The data presented show evidence for the presence of signalling compounds during AM symbiosis. While abundant data exist of the exact nature of certain signals during rhizobial symbiosis, no conclusive evidence exists concerning signalling in mycorrhizal association.

It is interesting to find that root exudates and SPC, such as flavonoids, found in root exudates can stimulate hyphal growth and are accumulated in mycorrhizal plants. Although some of these observations are reminiscent of the rhizobial symbiosis and thus the idea that they also might play an important role in AM is tempting, it can not be concluded at present that these compounds really have an essential function in the AM symbiosis.

SPC, such as flavonoids, have been studied mainly whether they are essential for the outcome of the symbiosis. The necessity of flavonoids for the establishment of AM has been previously questioned (Bécard et al., 1995). However, it is tempting to speculate that flavonoids, irrespective of their possible involvement during the formation of AM, play a role in mycorrhizal autoregulation and bioprotection.

It is surprising that the majority of studies on the bioprotective effect of mycorrhiza concentrate on the processes after root colonization and little information is available on the character of the changes occurring in the mycorrhizosphere. While defense mechanisms for bioprotection in the mycorrhizal root may be difficult to identify, identification of the mechanisms producing the changes in the mycorrhizosphere may be easier. Compounds in root exudates of mycorrhizal plants and/or extraradical mycelium of AMF, responsible for the observed effects should not be too difficult to identify. Key to these studies is an understanding of the mode of action of these compounds, involving the impact on the microbial population in the soil, including AMF.

In this work we extended current knowledge about signalling in AM symbiosis and compared it with signalling mechanisms in rhizobial symbiosis. We proposed some new hypotheses regarding the autoregulation of mycorrhization in connection with the bioprotective effects observed in mycorrhizal plants. Future studies will be focused on validating the hypotheses presented. Recent studies do show, however, exciting findings corroborating different aspects of these hypotheses.

7. ACKNOWLEDGMENTS

We thank Dr. C. Staehelin, Université de Genève, Switzerland for critical reading of the MS. This work was supported by a grant of the "Natural Sciences and Engineering Research Council of Canada".

8. REFERENCES

Albrecht, C., Geurts, R., Lapeyrie, F., and Bisseling, T., 1998, Endomycorrhizae and rhizobial nod factors both require SYM8 to induce the expression of the early nodulin genes *PsENOD5* and *PsENOD12A*, *Plant J.* **15**:605-614.

Azcon-Aguilar, C., and Barea, J. M., 1996, Arbuscular mycorrhizas and biological control of soil-borne plant pathogens - An overview of the mechanisms involved, *Mycorrhiza* **6**:457-464.

Bärtschi, H., Gianinazzi-Pearson, V., and Vegh, I., 1981, Vesicular-arbuscular mycorrhiza formation and root-rot disease (*Phytophthora cinnamomi* Rands) development in *Chamaecyparis lawsoniana* (Murr.) Parl, *Phytopathol. Z.* **102**:213-218.

Bansal, M., and Mukerji, K. G., 1994, Positive correlation between VAM-induced changes in root exudation and mycorrhizosphere mycoflora, *Mycorrhiza* **5**: 39-44.

Bécard, G., Douds, D. D., and Pfeffer, P. E., 1992, Extensive *in vitro* hyphal growth of vesicular-arbuscular mycorrhizal fungi in the presence of CO_2 and flavonols, *Appl. Environ. Microbiol.* **68**:1260-1264.

Bécard, G., Taylor, L. P., Douds, D. D., Pfeffer, P. E., and Doner, L. W., 1995, Flavonoids are not necessary plant signal compounds in arbuscular mycorrhizal symbiosis, *MPMI* **8**:252-258.

Blilou, I., Ocampo, J. A., and Garcia-Garrido, J. M., 1999, Resistance of pea roots to endomycorrhizal fungus or *Rhizobium* correlates with enhanced levels of endogenous salicylic acid, *J. Exp. Bot.* **50**:1663-1668.

Blilou, I., Ocampo, J. A., and Garcia-Garrido, J. M., 2000a, Induction of catalase and ascorbate peroxidase activities in tobacco roots inoculated with arbuscular mycorrhizal *Glomus mosseae*, *Mycol. Res.* **104**:722-725.

Blilou, I., Ocampo, J. A., and Garcia-Garrido, J. M., 2000b, Induction of *Ltp* (lipid transfer protein) and *Pal* (phenylalanine ammonia-lyase) gene expression in rice roots colonized by the arbuscular mycorrhizal fungus *Glomus mosseae*, *J. Exp. Bot.* **51**:1969-1977.

Bolanos-Vasquez, M. C., and Werner, D., 1997, Effect of *Rhizobium tropici*, *R. etli*, and *R. leguminosarum* bv. *phaseoli* on nod gene-inducing flavonoids in root exudates of *Phaseolus vulgaris*, *MPMI* **10**:339-346.

Caetano-Anollés, G., and Gresshof, P. M., 1991, Plant genetic control of nodulation, *Ann. Rev. Microbiol.* **45**:345-382.

Caron, M., Fortin J. A., and Richard, C., 1986, Effect of phosphorus concentration and *Glomus intraradices* on Fusarium crown and root rot of tomatoes, *Phytopathol.* **76**:942-946.

Chabot, S., Bel-Rhlid, R., Chênevert, R., and Piché, Y., 1992, Hyphal growth promotion *in vitro* of the VA mycorrhizal fungus, *Gigaspora margarita* Becker & Hall, by the activity of structurally specific flavonoids compounds under CO_2-enriched conditions, *New Phytol.* **122**:461-467.

Chakraborty, U., and Purkayastha, R. P., 1984, Role of rhizobitoxine in protecting soybean roots from *Macrophomina phaseolina* infection, *Can. J. Microbiol.* **30**:285-289.

Chakraborty, U., and Chakraborty, B. N., 1989, Interaction of *Rhizobium leguminosarum* and *Fusarium solani* f.sp. *pisi* on pea affecting disease development and phytoalexin production, *Can. J. Bot.* **67**:1698-1701.

Chaturvedi, C., and Singh, R., 1989, Response of chickpea (*Cicer arietinum* L.) to inoculation with *Rhizobium* and VA mycorrhiza, *Proc. Natl. Acad. Sci. India Sect. B* **59**:443-446.

Cluett, H. C., and Boucher, D. H., 1983, Indirect mutualism in the legume-*Rhizobium*-mycorrhizal fungus interaction, *Oecologia* **59**:405-408.

Cordier, C., Gianinazzi, S., and Gianinazzi-Pearson, V., 1996, Colonization patterns of root tissues by *Phytphthora nicotianae* var. *parasitica* related to reduced disease in mycorrhizal tomato, *Plant and Soil* **185**:223-232.

Cordier, C., Pozo, M. J., Barea, J. M., Gianinazzi, S., and Gianinazzi-Pearson, V., 1998, Cell defense responses associated with localized and systemic resistance to *Phytophthora parasitica* induced in tomato by an arbuscular mycorrhizal fungus, *MPMI* **11**:1017-1028.

Daft, M. J., and El-Giahmi, A. A., 1974, Effect of *Endogone* mycorrhiza on plant growth. VII. Influence of infection on the growth and nodulation in French bean (*Phaseolus vulgaris*), *New Phytol.* **73**:1139-1147.

Daft, M. J., and Hogarth, B. G., 1983, Competitive interactions of four species of *Glomus* on maize and onion, *Trans. Br. mycol. Soc.* **80**:339-345.

Dakora, F. D., Joseph, C. M., and Phillips, D. A., 1993, Alfalfa (*Medicago sativa* L.) root exudates contain isoflavonoids in the presence of *Rhizobium meliloti, Plant Physiol.* **101**:819-824.

Dakora, F. D., and Phillips, D. A., 1996, Diverse functions of isoflavonoids in legumes transcend anti-microbial definitions of phytoalexins, *Physiol. Mol. Plant Pathol.* **49**:1-20.

Davis, R. M., and Menge, J. A., 1980, Influence of *Glomus fasciculatus* and soil phosphorus on *Phytophthora* root rot of citrus, *Phytopathol.* **70**: 447-452.

Dehne, H. W., 1982, Interactions between vesicular-arbuscular mycorrhizal fungi and plant pathogens, *Phytopathol.* **72**:1115-1119.

Fester, T., Maier, W., and Strack, D., 1999, Accumulation of secondary compounds in barley and wheat roots in response to inoculation with arbuscular mycorrhizal fungi and co-inoculation with rhizosphere bacteria, *Mycorrhiza* **8**:241-246.

Feugey, L., Strullu, D. G., Poupard, P., and Simoneau, P., 1999, Induced defence responses limit Hartig net formation in ectomycorrhizal birch roots, *New Phytol.* **144**:541-547.

Fries, L. L., Pacovsky, R. S., Safir, G. R., and Siqueira, J. O., 1997, Plant growth and arbuscular mycorrhizal fungal colonization affected by exogenously applied phenolic compounds, *J. Chem. Ecol.* **23**:1755-1767.

Gaffney, T., Friedrich, L., Vernooij, B., Negrotto, D., Nye, G., Uknes, S., Ward, E., Kessmann, H., and Ryals, J., 1993, Requirement of salicylic acid for the induction of systemic required resistance, *Science* **261**:754-756.

Gemma, L. N., and Koske, R. E., 1988, Pre-infection interactions between roots and the mycorrhizal fungus *Gigaspora gigantea*: chemotropism of germ-tubes and root growth response, *Trans. Br. mycol. Soc.* **91**:123-132.

Gianinazzi-Pearson, V., Branzanti, B., and Gianinazzi, S., 1989, *In vitro* enhancement of spore germination and early hyphal growth of a vesicular-arbuscular mycorrhizal fungus by host root exudates and plant flavonoids, *Symbiosis* **7**:243-255.

Harley, J. L., and Harley, E. L., 1987, A check-list of mycorrhiza in the British flora, *New Phytol.* **105**:1-102.

Harrison, M. J., and Dixon, R. A., 1993, Isoflavonoid accumulation and expression of defense gene transcripts during the establishment of vesicular-arbuscular mycorrhizal associations in roots of *Medicago truncatula*, *MPMI* **6**:643-654.

Hepper, C. M., Azcon-Aguilar, C., Rosendahl, S., and Sen, R., 1988, Competition between three species of Glomus used as spatially separated introduced and indigenous mycorrhizal inocula for leek (*Allium porrum* L.), *New Phytol.* **110**:207-215.

Hirsch, A. M., and Kapulnik, Y., 1998, Signal transduction pathways in mycorrhizal associations: Comparisons with the *Rhizobium*-legume symbiosis, *Fung. Gen. Biol.* **23**:205-212.

Higgins, V. J., 1978, The effect of some pterocarpanoid phytoalexins on germ tube elongation of *Stemphylum botryosum*, *Phytopathol.* **68**:339-345.

Hooker, J. E., Jaizme-Vega, M., and Atkinson, D., 1994, Biocontrol of plant pathogens using arbuscular mycorrhizal fungi, In: *Impact of arbuscular mycorrhizas on sustainable agriculture and natural ecosystems*, Gianinazzi, S., and Schüepp, H., eds., Birkhäuser Verlag, Basel, Switzerland, pp. 191-200.

Kawai, Y., and Yamamoto, Y., 1986, Increase in the formation and nitrogen fixation of soybean nodules by vesicular-arbuscular mycorrhiza, *Plant Cell Physiol.* **27**:399-405.

Koske, R. E., 1982, Evidence for a volatile attractant from plant roots affecting germ tubes of a VA mycorrhizal fungus, *Trans. Br. mycol. Soc.* **79**:305-310.

Koske, R. E., and Gemma, J. N., 1992, Fungal reactions to plants prior to mycorrhizal formation, In: *Mycorrhizal functioning: An integrative plant fungal process*, Allen , M. F., ed., Chapman and Hall, New York, USA pp. 3-27.

Ludwig-Müller, J., 2000, Hormonal balance in plants during colonization by mycorrhizal fungi, In: *Arbuscular Mycorrhizas: Physiology and Function*, Kapulnik Y., and Douds, D. D., eds., Kluwer Academic Publishers, Dordrecht, The Netherlands, pp. 263-285.

Maier, W., Peipp, H., Schmidt, J., Wray, V., and Strack, D., 1995, Levels of a terpenoid glycoside (blumenin) and cell wall-bound phenolics in some cereal mycorrhizas, *Plant Physiol.* **109**:465-470.

Maier, W., Hammer, K., Dammann, U., Schulz, B., and Strack, D., 1997, Accumulation of sesquiterpenoid cyclohexenone derivatives induced by an arbuscular mycorrhizal fungus in members of the *Poaceae*, *Planta* **202**:36-42.

Maier, W., Schmidt, J., Wray, V., Walter, M. H., and Strack, D., 1999, The arbuscular mycorrhizal fungus, *Glomus intraradices*, induces the accumulation of cyclohexenone derivatives in tobacco roots, *Planta* **207**:620-623.

Maier, W., Schmidt, J., Nimtz, M., Wray, V., and Strack, D., 2000, Secondary products in mycorrhizal roots of tobacco and tomato, *Phytochem.* **54**:473-479.

Malamy, J., and Klessig, D. F., 1992, Salicylic acid and plant disease resistance, *Plant J.* 2:643-654.

Meyer, J. R., and Linderman, R. G., 1986, Selective influence in populations of rhizosphere or rhizoplane bacteria and actinomycetes by mycorrhizas formed by *Glomus fasciculatum*, *Soil Biol. Biochem.* 18:191-196.

Miersch, O., and Wasternack, C., 2000, Octadecanoid and jasmonate signaling in tomato (*Lycopersicon esculentum* Mill.) leaves: Endogenous jasmonates do not induce jasmonate biosynthesis, *Biol. Chem.* 381:715-722.

Morandi, D., 1996, Occurence of phytoalexins and phenolic compounds on endomycorrhizal interactions, and their potential role in biological control, *Plant and Soil* 185:241-251.

Norman, J. R., and Hooker, J. E., 2000, Sporulation of *Phytophthora fragariae* shows greater stimulation by exudates of non-mycorrhizal than by mycorrhizal strawberry roots, *Mycol. Res.* 104:1069-1073.

Pacovsky, R. S., Fuller, G., Stafford, A. E., and Paul, E. A., 1986, Nutrient and growth interaction in soybeans colonized with *Glomus fasciculatum* and *Rhizobium japonicum*, *Plant and Soil* 92:37-45.

Pearson, J. N., Abbott, L. K., and Jasper D. A., 1993, Mediation of competition between two colonizing VA mycorrhizal fungi by the host plant, *New Phytol.* 123:93-98.

Pearson, J. N., Abbott, L. K., and Jasper, D. A., 1994, Phosporus, soluble carbohydrates and the competition between two arbuscular mycorrhizal fungi colonizing subterranean clover, *New Phytol.* 127:101-106.

Peipp, H., Maier, W., Schmidt, J., Wray, V., and Strack, D., 1997, Arbuscular mycorrhizal fungus-induced changes in the accumulation of secondary compounds in barley roots, *Phytochem.* 44:581-587.

Perret, X., Staehelin, C., and Broughton, W. J., 2000, Molecular basis of symbiotic promiscuity, *Microbiol. Mol. Biol. Rev.* 64:180-201.

Phillips, D. A., and Tsai, S. M., 1992, Flavonoids as plant signals to the rhizosphere microbes, *Mycorrhiza* 1:55-58.

Pieterse, C. M. J., and van Loon, L. C., 1999, Salicylic acid-independent plant defense pathways, *Trends in Plant Science* 4:52-58.

Pinior, A., Wyss, U., Piché, Y., and Vierheilig H., 1999, Plants colonized by AM fungi regulate further root colonization by AM fungi through altered root exudation, *Can. J. Bot.* 77:891-897.

Poulin, M.-J., Bel-Rhlid, R., Piché, Y., and Chênevert, R., 1993, Flavonoids released by carrot (*Daucus carota*) seedlings stimulate hyphal development of vesicular-arbuscular mycorrhizal fungi in the presence of optimal CO_2 enrichment, *J. Chem. Ecol.* 19:2317-2327.

Rosendahl, S., 1985, Interactions between the vesicular-arbuscular mycorrhizal fungus *Glomus fasciculatum* and *Aphanomyces euteiches* root rot of peas, *Phytopathol. Z.* 114:31-40.

Recourt, K., Van Tunen, A. J., Mur, L. A., Van Brussel, A. A. N., Lugtenberg, B., and Kijne, J. W., 1992, Activation of flavonoid biosynthesis in roots of *Vicia sativa* subsp. *nigra* plants by inoculation with *Rhizobium leguminosarum* biovar *viciae*, *Plant Mol. Biol.* 19:411-420.

Schmidt, P. E., Broughton, W. J., and Werner, D., 1994, Nod factors of *Bradyrhizobium japonicum* and *Rhizobium* sp. NGR234 induce flavonoid accumulation in soybean root exudates, *MPMI* 7:384-390.

Slezack, S., Dumas-Gaudot, E., Paynot, M., and Gianinazzi, S., 2000, Is a fully established arbuscular mycorrhizal symbiosis required for bioprotection of *Pisum sativum* roots against *Aphanomyces euteiches* ? *MPMI* 13: 238-241.

Smith, S., and Read, D. J., 1997, *Mycorrhizal Symbiosis*, Academic Press, London, UK.

St.-Arnaud, M., Hamel C., Vimard, B., Caron, M., and Fortin, J. A., 1997, Inhibition of *Fusarium oxysporum* f.sp. *dianthi* in the non-VAM species *Dianthus caryophyllus* by co-culture with *Tagetes patula* companion plants colonized by *Glomus intraradices*, *Can. J. Bot.* 75:998-1005.

St.-Arnaud, M., Hamel, C., Caron, M., and Fortin, J. A., 1995, Endomycorhizes VA et sensibilité des plantes aux maladie: synthèse de la littérature et mécanismes d'interaction potentiels, In: *La symbiose mycorhizienne*, Fortin, J. A., Charest, C., and Piché, Y., eds., Éditions Orbis, Frelighsburg, Québec, Canada pp. 51-87.

Sticher, L, Mauch-Mani, B., and Métraux, J. P., 1997, Systemic acquired resistance, *Ann. Rev. Phytopathol.* 35:235-270.

Suriyapperuma, S. P., and Koske, R. E., 1995, Attraction of germ tubes and germination of spores of the arbuscular mycorrhizal fungus *Gigaspora gigantea* in the presence of roots of maize exposed to different concentrations of phosphorus, *Mycologia* 87:772-778.

Tawaraya, K., Hashimoto, K., and Wagatsuma, T., 1998, Effect of root exudate fractions from P-deficient and P-sufficient onion plants on root colonisation by the arbuscular mycorrhizal fungus *Gigaspora margarita*, *Mycorrhiza* 8:67-70.

Tu, J. C., 1978, Protection of soybean from severe *Phytophthora* root rot by *Rhizobium*, *Physiol. Plant Pathol.* 12:233-240.

Van Etten, H. D., 1976, Antifungal activity of pterocarpans and other selected isoflavonoids, *Phytochem.* 15:655-659.

Vierheilig, H., Alt, M., Mohr, U., Boller, T., and Wiemken, A., 1994, Ethylene biosynthesis and activities of chitinase and ß-1,3-glucanase in the roots of host and non-host plants of vesicular-arbuscular mycorrhizal fungi after inoculation with *Glomus mosseae*, *J. Plant Physiol.* 143:337-343.

Vierheilig, H., Bago, B., Albrecht, C., Poulin, M.-P., and Piché, Y., 1998a, Flavonoids and arbuscular-mycorrhizal fungi, In: *Flavonoids in the Living System*, Manthey, J. A., and Buslig, B. S., eds., Plenum Press, New York, USA, pp. 9-33.

Vierheilig, H., Alt-Hug, M., Engel-Streitwolf, R., Mäder, P., and Wiemken, A., 1998b, Studies on the attractional effect of root exudates on hyphal growth of an arbuscular mycorrhizal fungus in a soil compartment-membrane system, *Plant and Soil* **203**:137-144.

Vierheilig, H., Coughlan, A.P., Wyss, U., and Piché, Y., 1998c, Ink and vinegar, a simple staining technique for arbuscular-mycorrhizal fungi, *Appl. Environ. Microbiol.* **64**:5004-5007.

Vierheilig, H., Bennett, R., Kiddle, G., Kaldorf, M., and Ludwig-Müller, J., 2000a, Differences in glucosinolate patterns and arbuscular mycorrhizal status of glucosinolate-containing plant species, *New Phytol.* **156**:343-352.

Vierheilig, H., Gagnon, H., Strack, D., and Maier, W., 2000b, Accumulation of cyclohexenone derivatives in barley, wheat and maize roots in response to inoculation with different arbuscular mycorrhizal fungi, *Mycorrhiza* **9**:291-293.

Vierheilig, H., Garcia-Garrido, J. M., Wyss, U., and Piché, Y., 2000c, Systemic suppression of mycorrhizal colonization of barley roots already colonized by AM fungi, *Soil Biol. Biochem.* **32**:589-595.

Vierheilig, H., Maier, W., Wyss, U., Samson, J., Strack, D., and Piché, Y., 2000d, Cyclohexenone derivative- and phosphate-levels in split-root systems and their role in the systemic suppression of mycorrhization in precolonized barley plants, *J. Plant Physiol.* **157**:593-599.

Vigo, C., Norman, J. R., and Hooker, J. E., 2000, Biocontrol of the pathogen *Phytophthora parasitica* by arbuscular mycorrhizal fungi is a consequence of effects on infection loci, *Plant Pathol.* **49**:509-514.

Volpin, H., Elkind, Y., Okon, Y., and Kapulnik, Y. A., 1994, vesicular-arbuscular mycorrhizal fungus (*Glomus intraradix*) induces a defense response in alfalfa roots, *Plant Physiol.* **104**:683-689.

Weidenbörner, M., and Jha H. C., 1994, Structure-activity relationship among isoflavonoids with regard to their antifungal properties, *Mycol. Res.* **98**:1376-1378.

Wilson, J. M., 1984, Competition for infection between vesicular-arbuscular mycorrhizal fungi, *New Phytol.* **97**:427-435.

Wyman, J. G., and Van Etten, H. D., 1978, Antibacterial activity of selected isoflavonoids, *Phytopathol.* **68**:583-589.

Xie, Z-P., Staehelin, C., Vierheilig, H., Wiemken, A., Jabbouri, S., Broughton, W. J., Vögeli-Lange, R., and Boller, T., 1995, Rhizobial nodulation factors stimulate mycorrhizal colonization of nodulating and non nodulating soybeans, *Plant Physiol.* **108**:1519-1525.

Xie, Z-P., Müller, J., Wiemken, A., Broughton, W. J., and Boller, T., 1998, Nod factor and *tri*-iodobenzoic acid stimulate mycorrhizal colonization and affect carbohydrate partitioning in mycorrhizal roots of *Lablab purpureus*, *New Phytol.* **139**:361-366.

Xie, Z-P., Staehelin, C., Wiemken, A., Broughton, W. J., Müller, J., and Boller, T., 1999, Symbiosis-stimulated chitinase isoenzymes of soybean (*Glycine max* (L.) Merr.), *J. Exp. Bot.* **50**:327-333.

THE USE OF A PHOTOACTIVATABLE KAEMPFEROL ANALOGUE TO PROBE THE ROLE OF FLAVONOL 3-*O*-GALACTOSYLTRANSFERASE IN POLLEN GERMINATION

Loverine P. Taylor[1] and Keith D. Miller

1. ABSTRACT

Flavonol induced pollen germination in petunia is rapid, specific, and achieved at low concentrations of kaempferol or quercetin. To determine the macromolecules that interact with the flavonol signal we have synthesized affinity-tagged kaempferol analogues. The first generation molecules are based on a benzophenone photophore. We find that 2-(3-benzoylphenyl)-3,5,7-trihydroxychromen-4-one (BPKae) antagonizes flavonol-induced pollen germination in a concentration-dependent manner. Further, BPKae acts as an irreversible inhibitor of flavonol 3-*O*-galactosyltransferase (F3GalTase), the gametophyte-specific enzyme that controls the accumulation of glycosylated flavonols in pollen. The effects of BPKae are mediated by UV-A light treatment. The binding characteristics of BPKae to F3GalTase suggest that it can be used to identify the residues required for flavonol-binding and catalysis.

2. BACKGROUND

2.1. Flavonols are Required for Pollen Germination and Tube Growth

When pollen falls on a receptive stigma, the stored RNA, protein, and bioactive small molecules allow rapid germination and outgrowth of a tube that delivers the sperm cells to the embryo sac to initiate fertilization (Taylor and Hepler, 1997). Although pollination is

[1] School of Molecular Biosciences, Washington State University, Pullman WA 99164-4234; Phone: 509-335-3612; FAX: 509-335-1907; e-Mail: LTAYLOR@wsu.edu

Flavonoids in Cell Function, edited by Béla Buslig and John Manthey.
Kluwer Academic/Plenum Publishers, New York, 2002.

41

fundamental to reproductive success, few of the molecules involved in the process have been identified. One exception is the requirement for flavonols in pollen germination and tube growth (Taylor and Hepler, 1997).

In maize and petunia, flavonol-deficient pollen is Conditionally Male Fertile (CMF); although it is viable it does not germinate or produce a functional tube unless supplied with an exogenous source of flavonol (Mo et al., 1992; Taylor and Jorgensen, 1992). Kaempferol is the most potent inducer of pollen germination: at 0.4 μM it induces complete germination of an *in vitro* suspension of CMF pollen (pollen rescue) (Mo et al., 1992; Vogt et al., 1995; Taylor et al., 1998). Metabolic studies of CMF pollen treated with a [^{14}C]kaempferol derivative (Xu et al., 1997) showed that the flavonol molecule was not catabolized during germination but was rapidly and quantitatively conjugated to specific sugars forming a single, pollen-specific class of flavonol 3-*O*-diglycoside (Vogt and Taylor, 1995; Xu et al., 1997). Hydration studies have detected tube outgrowth within 2-3 min of flavonol addition, virtually the same kinetics as wild-type pollen (Taylor et al., 1998). Conversion of the added aglycone to the glycosylated form is even more rapid; the flavonol 3-*O*-disaccharide is detected within 1 min of flavonol addition (Vogt and Taylor, 1995; Xu et al., 1997; Taylor et al., 1998). Thus flavonol activity in germination may be controlled by enzymes that regulate the glycosylation level of flavonols.

2.2. A Gametophyte-Specific Flavonol 3-*O*-Galactosyltransferase Controls the Accumulation of Flavonols in Pollen

Petunia pollen accumulates two unique flavonol glycosides, kaempferol and quercetin-3-*O*-(2"-*O*-β-D-glucopyranosyl)-β-D-galactopyranoside (Zerback et al., 1989; Vogt and Taylor, 1995). We have isolated and characterized the pollen-specific enzyme that controls the accumulation of flavonol glycosides (Vogt and Taylor, 1995; Miller et al., 1999). Unlike the relatively non-specific substrate usage of flavonoid glycosyltransferases from sporophytic tissues, the pollen flavonol 3-*O*-galactosyltransferase (F3GalTase) is a highly discriminating and efficient enzyme. F3GalTase rapidly conjugated the flavonols that induce pollen germination to UDP-galactose, forming the 3-*O*-galactosides. Heterologous expression of the F3GalTase cDNA in *E. coli* yielded active recombinant enzyme (rF3GalTase) which had the identical substrate specificity as the native enzyme. Kinetic analysis showed that the k_{cat}/K_m values of rF3GalTase, using kaempferol and quercetin as substrates, approached that of a catalytically perfect enzyme (Miller et al., 1999). We previously showed that in addition to flavonol aglycones, two flavonol glycosides could stimulate pollen germination. The active molecules were kaempferol and quercetin 3-*O*-galactoside, the endogenous products of F3GalTase activity in pollen. No other flavonol glycosides were active (Taylor et al., 1998). In the same study we provided evidence that the ability of kaempferol and quercetin 3-*O*-galactoside to induce pollen germination was mediated by the reverse activity of F3GalTase. Kinetic studies of the purified rF3GalTase showed that it catalyzed the reverse reaction, generation of flavonols from UDP and flavonol 3-*O*-galactosides, almost as efficiently as the forward reaction (Miller et al., 1999). The forward and reverse activities of F3GalTase showed pH optima of 8.0 and 6.0, respectively, suggesting that one reaction might be favored over another in various subcellular compartments. The final step, formation of the diglycoside by the addition of glucose in a 1→2 configuration to the galactoside, was irreversible and produced a molecule which could not stimulate germination *in vitro* (Taylor et al., 1998). Importantly, formation of the flavonol 3-*O*-galactoside was a prerequisite to formation of the flavonol diglycoside (Vogt and Taylor, 1995; Miller et al., 1999).

2.3. Flavonol Synthesis and Metabolism in Anthers: Regulation by Compartmentalization

Virtually all pollens accumulate flavonols, often to very high levels (>2% dry weight) (Taylor and Hepler, 1997). All of the current biochemical and histological evidence indicate that flavonols function only at germination; no deleterious effects of flavonol deficiency have been detected during pollen development (Taylor and Jorgensen, 1992; Pollak et al., 1993). Pollen flavonols are synthesized in a complex multistep pathway that requires temporal and spatial coordination of parallel developmental programs in the sporophyte (anther) and the gametophyte (pollen). The result is the sequestration in the pollen of unique molecules that are required for germination and tube growth. Briefly, flavonol aglycones (e. g. kaempferol) are synthesized by enzymes active in the tapetal layer of the anther wall. The aglycones are released into the anther locule by diffusion or upon rupture of the tapetal cells when they undergo a programmed death. Kaempferol molecules (K in Fig.1) in the locule diffuse or are transported into the developing pollen grain where they

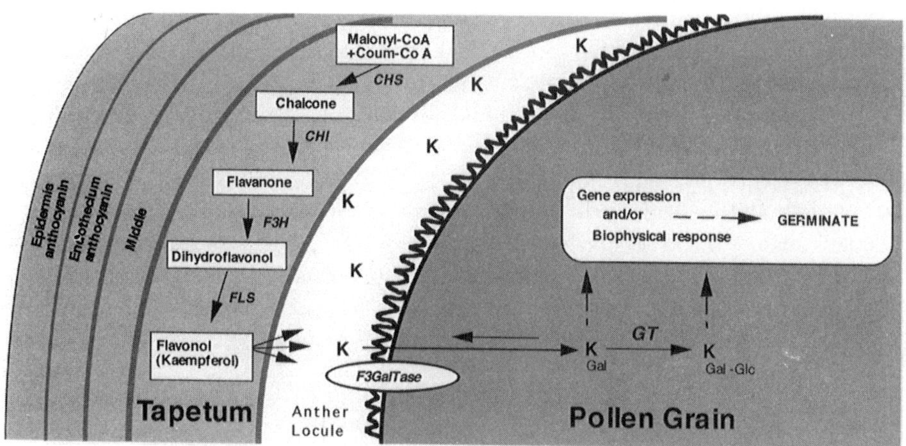

Figure 1. Flavonol biosynthesis and metabolism in developing anthers and pollen.

are modified to pollen-specific glycosides by the action of gametophytically-encoded glycosyltransferases. Conjugation to sugars renders the molecules water soluble and may provide a way to sequester the 'spent' flavonol signal. Alternatively the galactoside and/or the diglycoside may function as a latent pool of potentially active molecules. The reverse activity of F3GalTase provides a mechanism for the galactoside to be converted to the active aglycone. On the other hand, despite intense efforts to detect an activity, we have found no evidence of a specific glycosidase that generates aglycones from the diglycosides.

2.4. Experimental approach

Using a series of naturally occurring and chemically modified compounds, a detailed structure-activity analysis identified the sites on the flavonol molecule that were required for germination-inducing activity *in vitro* (Mo et al., 1992; Vogt et al., 1995; Taylor et al., 1998). This information was critical for the design of bioactive affinity ligands. Briefly, for ring A/C there is an absolute requirement for the 2,3-double bond and the hydroxyl groups at carbons 3, 5 and 7. Substituents at C-6 and C-8 (OH, OMe) are tolerated. For ring B the structural requirements indicate a hydrophobic interaction which is perturbed by the addition of polar hydroxyl groups. Presumably this reflects the efficiency with which the added compounds are taken up and metabolized by intact pollen grain.

In the present study three assays were used to characterize the effect of the flavonol affinity tag on pollen germination and interacting macromolecules. The pollen rescue assay is a simple and defined model system which we use to test the effect of various treatments on pollen germination and flavonol metabolism. In its basic form it involves the addition of 1 µM kaempferol (or other compounds) to a suspension of flavonol-deficient CMF pollen in germination medium (GM). Activity is measured in terms of the number of pollen grains germinated after 2h incubation (%G). The metabolic fate of the flavonol (or other molecules) in the treated pollen was determined by HPLC and spectral analysis of an extract of the germinating pollen suspension (Mo et al., 1992; Vogt and Taylor, 1995; Xu et al., 1997; Taylor et al., 1998). F3GalTase activity was determined using the standard assay conditions which include 2.5 ng of purified recombinant enzyme and 25 µM kaempferol and 5 mM UDPgalactose as substrates (Miller et al., 1999).

The specificity, sensitivity and rapidity of the germination response to flavonols suggests a signaling role. To elucidate the underlying molecular mechanism of flavonol-induced pollen germination it is necessary to identify the specific proteins that interact with, and transduce, the flavonol signal in pollen. Unfortunately biochemical evidence of specific flavonoid binding is scarce. Flavonol interaction with *Rhizobium* gene products has been inferred by genetic studies but not corroborated by biochemical evidence (Györgypal et al., 1991). In plants, a putative flavonol sulfate binding protein was detected in nuclear extracts

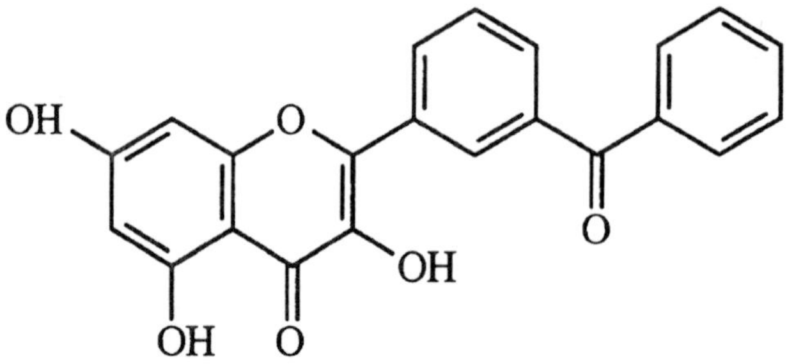

Figure 2. 2-(3-benzoylphenyl)-3,5,7-trihydroxychromen-4-one (BPKae).

from leaves using an antibody to a flavonol sulfate-BSA conjugate (Grandmaison and Ibrahim, 1996)). However technical difficulties such as non-specific binding (phenolics are notorious for this), low specific activity probes and leaching of flavonols from the disrupted tissue makes the demonstration of specific interactions difficult. To circumvent these problems we have embarked on a program to develop and synthesize a series of high affinity flavonol probes which can be cross-linked to flavonol-binding macromolecules.

3. RESULTS

3.1. Synthesis of a Photoactivatable Kaempferol Analogue

Affinity labeling of a macromolecule involves the covalent introduction of chemical tags attached to an interacting ligand. Photoaffinity labeling is a variation whereby the reactive group is inert in the dark but can be converted to a highly reactive intermediate upon irradiation by a specific wavelength of light (Ruoho et al., 1973; Chowdhry and Westheimer, 1979). Based on the structure-activity analyses, we concluded that molecules with chemical modifications to the B ring had the greatest probability to retain biological activity. Therefore the first molecule we synthesized was the 4'-benzoyl derivative of kaempferol (Fig. 2)(Tanaka et al., 2000). The benzophenone (BP) chromophore has several attractive features as a photoprobe: BP's are chemically stable and can be manipulated in ambient light; the BP photophore is activated at 350-360 nm (UV-A), avoiding protein-damaging wavelengths; and BP's react preferentially with unreactive C-H bonds even in the presence of solvent water and bulk nucleophiles (Dormán and Prestwich, 1994). Details of the synthesis have been published (Tanaka et al., 2000), and the key step involved a Robinson's Annulation Reaction (Allan and Robinson, 1924). ω-Benzyloxy-acetophenone was condensed with 3-benzoyl-benzoic acid at moderate temperature and the subsequent dehydration step led to irreversible ring closure. The identity and purity of the BPKae product was ascertained by mass spectroscopy (LC-MS) and nuclear magnetic resonance spectroscopy (NMR).

3.2. BPKae Antagonizes Pollen Germination in a Concentration-Dependent Manner

The usefulness of a flavonol affinity tag depends on how closely it mimics the physiological and chemical properties of the endogenous kaempferol ligand. This cannot be predicted *a priori* so each molecule must be tested empirically. BPKae was unable to induce CMF pollen to germinate, however it did antagonize kaempferol-induced CMF pollen germination in a concentration- and UV-A-dependent manner. A suspension of CMF pollen in GM was treated with increasing concentrations of BPKae and exposed to 20 min of UV-A light. The pollen suspension was then treated with 1 μM kaempferol to induce germination. As shown in Fig. 3, 25 μM BPKae caused an almost 85% reduction in CMF pollen germination frequency. The effect did not occur in the absence of UV-A treatment (data not shown). As a control, wild type pollen was treated with BPKae and UV-A with no toxic effect on germination noted at concentrations up to 100μM.

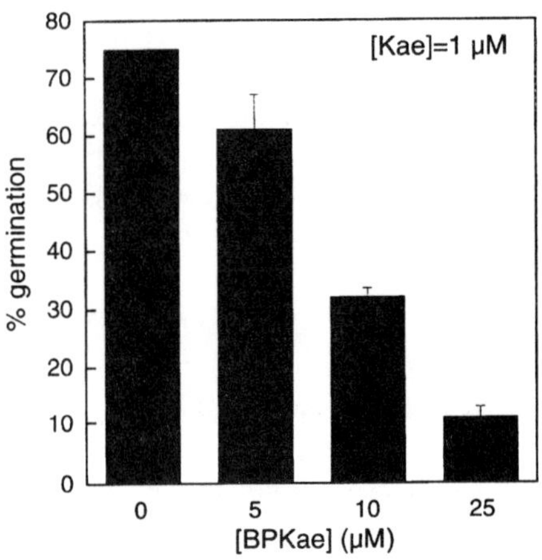

Figure 3. Pollen germination rescue by kaempferol is reduced by BPKae and UV-A light treatment.

3.3. BPKae is a Substrate for Flavonol 3-*O*-Galactosyltransferase

The macromolecules responsible for detecting and transducing the flavonol signal during pollen germination are unknown. However the identity of one flavonol-binding protein in germinating pollen is known: F3GalTase binds the kaempferol molecule with a K_m of 1.1 µM and rapidly converts it to the 3-*O*-galactoside. To determine if F3GalTase might be a target for the antagonizing influence of BPKae on germination, we characterized the effect of BPKae and UV-A on F3GalTase activity. We determined that BPKae functions as a substrate for F3GalTase, albeit at reduced efficiency when compared to the native substrate kaempferol (about 20-30% as active). When 25 µM BPKae was substituted for kaempferol in the standard F3GalTase assay, a single product was detected by HPLC analysis of the reaction mixture (data not shown). The product was determined to be the galactoside of BPKae by retention time, spectral profile and LC-MS. This result suggests that BPKae binds F3GalTase in the catalytic site.

3.4. BPKae is an Irreversible Inhibitor of F3GalTase Activity

Although BPKae acts as a substrate for F3GalTase, it will not be effective as a photo-affinity probe unless it can be cross-linked to the target protein. To determine the require-ment for UV-A-induced cross-linking of BPKae to F3GalTase, the standard F3GalTase assay was modified by the addition of 25 µM BPKae and subjected to increasing length UV-A light exposure. Fig. 4 shows that even 10 min of UV-A treatment caused a reduction in rF3GalTase activity. After 60 min of exposure, formation of kaempferol 3-*O*-galactoside was reduced by 90% relative to a control which consisted of UV-A treated enzyme in the absence of BPKae. This result shows a direct correlation between increasing time of UV-A treatment and an increase in BPKae-mediated inhibition of rF3GalTase activity. The finding that BPKae cross-links F3GalTase and inhibits activity indicates that the

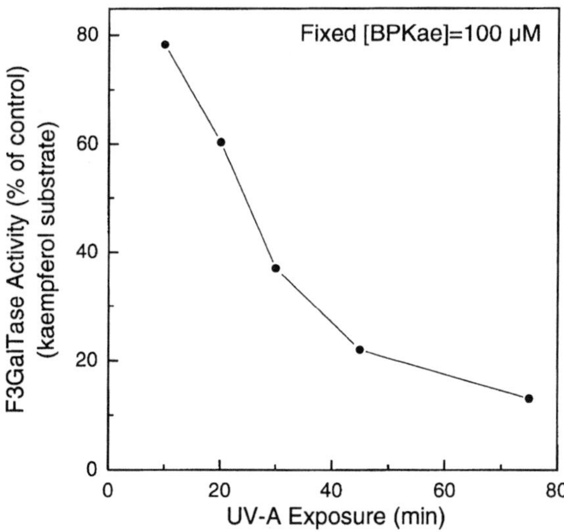

Figure 4. UV-A induced photolysis is required for maximal inhibition of flavonol 3-*O*-galactosyltransferase activity by BPKae.

kaempferol-derivative can be used to probe the flavonol-binding site. However, interactions between BPKae and F3GalTase must conform to certain diagnostic features. First, binding to a single site must be demonstrated (F3GalTase is a non-allosteric, monomeric enzyme). Second, the endogenous ligand kaempferol should protect F3GalTase against inhibition by BPKae. To confirm that BPKae binds F3GalTase at a single site, the effect of increasing amounts of BPKae on rF3GalTase activity were determined. As shown in Fig. 5, BPKae binding is saturable which indicates a single binding site on the F3GalTase molecule. To confirm that this was the same site used by the endogenous substrate kaempferol (i.e. the catalytic site) F3GalTase activity was determined in the presence 25 µM of BPKae and increasing amounts of kaempferol. As shown in Fig. 6, 25 µM kaempferol could completely abolish the inhibitory effects of BPKae on F3GalTase activity. The saturable binding of BPKae to F3GalTase and the protective effects of the endogenous substrate kaempferol on the inhibition by BPKae indicates that BPKae inhibits F3GalTase by binding and cross-linking to the active site. Thus it will be a useful probe to determine the residues in F3GalTase necessary for flavonol-binding and catalysis.

4. DISCUSSION

A variety of proteins may be labeled with a flavonol affinity probe including receptors, transport proteins, and flavonol biosynthetic enzymes. The latter class of flavonol-binding proteins have been well characterized in a number of species and yet to date no flavonol-binding site has ever been identified. The information presented here shows that important residues of F3GalTase critical for substrate binding or catalysis could be identified by

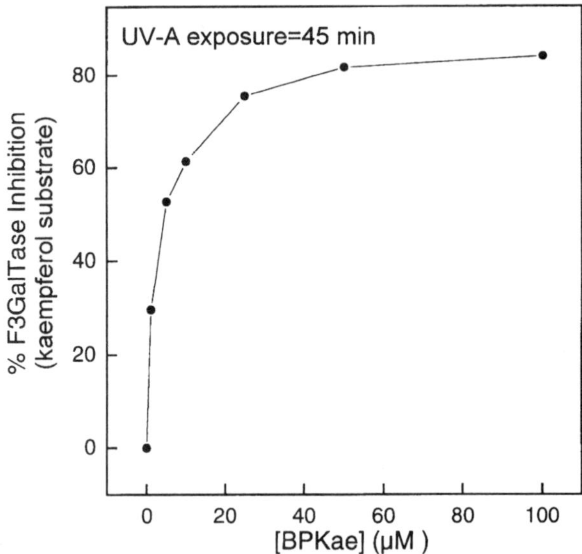

Figure 5. BPKae binding to F3GalTase is saturable.

radiolabeling BPKae and UV-A cross-linking. Biochemical studies of a number of flavonoid glycosyltransferases have shown that sulfhydryl binding reagents have a negative effect on activity suggesting that cysteine residue(s) are important for catalysis (Ishikura and Mato, 1993; Vogt et al., 1997). BPKae could be used to confirm this. Also BPKae could be used

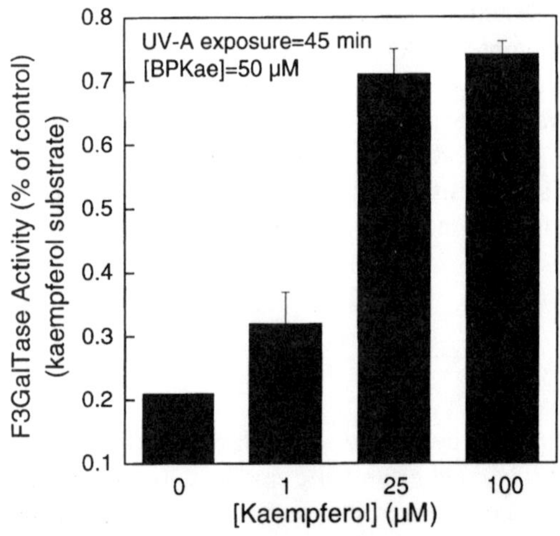

Figure 6. Kaempferol protects F3GalTase from inhibition by BPKae.

to determine the sequential order of multiple modifying activities and to probe the catalytic residues of biosynthetic enzymes such as flavonol synthase. In addition to providing the first details of a flavonol-binding site in a flavonoid glycosyltransferase, BPKae might determine a role for F3GalTase in pollen germination. An important issue to resolve is whether F3GalTase plays a direct role in flavonol-induced pollen germination or if it functions to metabolize a "spent" germination signal. Awareness is growing for the crucial importance of glycosylating activities in plant development. A recent report described a pea-meristem localized glucoronosyltransferase that was critical for proper tissue differentiation (Woo et al., 1999). Given that kaempferol galactoside formation precedes visible tube outgrowth, it is possible that F3GalTase itself could function as the receptor of the flavonol signal in petunia pollen. Alternatively, if flavonol 3-*O*-galactoside is the *in vivo* germination-inducing signal, then the reverse activity of F3GalTase could generate the active molecule. Demonstrating that a BPKae-mediated reduction in F3GalTase activity is accompanied by a decrease in pollen germination or tube growth would confirm an essential role for F3GalTase in plant fertility. To date we have been unable to establish this correlation which might reflect the complexities of uptake, compartmentation, and metabolism of BPKae by the intact pollen grain. On the other hand, the fact that BPKae does not stimulate pollen germination may limit its usefulness in this regard. The use of affinity-tagged molecules that induce germination and are also substrates for F3GalTase might be a better approach. Alternatively, molecules that induce germination and are not substrates for F3GalTase would quickly rule out a role for F3GalTase in germination. The second generation flavonols tagged with an azido group show promise for these approaches because they induce pollen germination.

5. REFERENCES

Allan, J., and Robinson, R., 1924, An accessible derivative of chromonol, *J. Chem. Soc.* **125**:2192-2195.

Chowdhry, V., and Westheimer, F. H., 1979, Photoaffinity labeling of biological systems, *Ann. Rev. Biochem.* **48**:293-325.

Dormán, G., and Prestwich, G. D., 1994, Benzophenone photophores in biochemistry, *Biochemistry* **33**:5661-5673.

Grandmaison, J., and Ibrahim, R. K., 1996, Evidence for nuclear protein binding of flavonol sulfate esters in *Flaveria cloraefolia*, *J. Plant Physiol.* **147**:653-660.

Györgypal, Z., Kiss, G. B., and Kondorosi, A., 1991, Transduction of plant signal molecules by the *Rhizobium NodD* proteins, *Bioessays* **13**:575-581.

Ishikura, N., and Mato, M., 1993, Partial purification and some properties of flavonol 3-*O*-glycosyltransferases from seedlings of *Vigna mungo* with special reference to the formation of kaempferol-3-*O*-galactoside and the 3-*O*-glucoside, *Plant Cell Physiol.* **34**:329-335.

Miller, K. D., Guyon, V., Evans, J. N. S., Shuttleworth, W. A., and Taylor, L. P., 1999, Purification, cloning, and heterologous expression of a catalytically efficient flavonol 3-*O*-galactosyltransferase expressed in the male gametophyte of *Petunia hybrida*, *J. Biol. Chem.* **274**:34011-34019.

Mo, Y., Nagel, C., and Taylor, L. P., 1992, Biochemical complementation of chalcone synthase mutants defines a role for flavonols in functional pollen, *Proc. Natl. Acad. Sci. USA* **89**:7213-7217.

Pollak, P. E., Vogt, T., Mo, Y., and Taylor, L. P., 1993, Chalcone synthase and flavonol accumulation in stigmas and anthers of *Petunia hybrida*, *Plant Physiol.* **102**:925-932.

Ruoho, A. E., Kiefer, H., Roeder, P. E., and Singer, S. J., 1973, The mechanism of photoaffinity labeling, *Proc. Natl. Acad. Sci. USA* **70**:2567-2571.

Tanaka, H., Stohlmeyer, M. M., Wandless, T. J., and Taylor, L. P., 2000, Synthesis of flavonol derivatives as probes of biological processes, *Tetrahedron Lett.* **41**:9735-9739.

Taylor, L. P., and Hepler, P. R., 1997, Pollen germination and tube growth, *Ann. Rev. Plant Phys. Plant Mol. Biol.* **48**:461-491.

Taylor, L. P., and Jorgensen, R., 1992, Conditional male fertility in chalcone synthase-deficient petunia, *J. of Hered.* **83**:11-17.

Taylor, L. P., Strenge, D., and Miller, K., 1998, The role of glycosylation in flavonol-induced pollen germination, In *Flavonoids in the Living System*, Manthey, J. A., and Buslig, B. S., eds., Plenum Press, New York pp. 35-44.

Vogt, T., and Taylor, L. P., 1995, Flavonol 3-*O*-glycosyltransferases associated with petunia pollen produce gametophyte-specific flavonol diglycosides, *Plant Physiol.* **108**:903-911.

Vogt, T., Wollenweber, E., and Taylor, L. P., 1995, The structural requirements of flavonols that induce pollen germination of conditionally male fertile *Petunia*, *Phytochemistry* **38**:589-592.

Vogt, T., Zimmermann, E., Grimm, R., Meyer, M., and Strack, D., 1997, Are the characteristic of betanidin glucosyltransferases from cell-suspension cultures of *Dorotheanthus bellidiformis* indicative of their phylogenic relationship with flavonoid glucosyltransferases? *Planta* **203**:349-361.

Woo, H.-H., Orbach, M. J., Hirsch, A. M., and Hawes, M. C., 1999, Meristem localized inducible expression of a UDP-glycosyltransferase gene is essential for growth and development in pea and alfalfa, *Plant Cell* **11**:2303-2315.

Xu, P., Vogt, T., and Taylor, L. P., 1997, Uptake and metabolism of flavonols during in vito germination and tube growth of petunia pollen, *Planta* **202**:257-265.

Zerback, R., Bokel, M., Geiger, H., and Hess, D., 1989, A kaempferol 3-glucosylgalactoside and further flavonoids from pollen of *Petunia hybrida*, *Phytochemistry* **28**:897-899.

FLAVONOIDS: SIGNAL MOLECULES
IN PLANT DEVELOPMENT

Ho-Hyung Woo[1,2], Gary Kuleck[3], Ann M. Hirsch[1,4], and Martha C. Hawes[2]

1. BACKGROUND

More than 4,000 different flavonoids, which are widely distributed in higher plants, have been identified (Harborne, 1988; Koes *et al.* 1994). Flavonoids are involved in numerous functions in vascular plants. One obvious function is as a filter for ultraviolet light, which requires that there be relatively large concentrations of flavonoids in epidermal cells (Li *et al.*, 1993). Flavonoids also function as external chemical signals for symbiotic nitrogen fixation (Fisher and Long, 1992; Baker, 1992) and for mycorrhizal relationships (Vierheilig *et al.*, 1998). Flavonoids are also important in fighting disease; ingress of pathogenic microbes can be curtailed in certain incompatible reactions in part because some plants synthesize flavonoid phytoalexins (Dixon and Paiva, 1995). Anthocyanidins and their glycosides, accumulating mainly in the inner epidermis of the petal just prior to opening of the flower bud, attract insect and other agents for pollination (Weiss, 1991). Later on, as the fruit develops, flavonoids can also attract agents for seed dispersal.

Flavonoids also act as internal physiological regulators or chemical messengers in the plant, and for these functions, relatively small amounts are effective. Flavonoids such as quercetin and kaempferol have been shown to function as auxin transport inhibitors (Jacobs and Rubery, 1988). Monohydroxy B-ring flavonoids are known to be involved in the degradation of the plant growth hormone, indole acetic acid, whereas dihydroxy B-ring flavonoids inhibits IAA degrading-activity (Furuya *et al.*, 1962; Galston, 1969). Flavonoids are also involved in pollen development (Mo *et al.*, 1992; van der Meer *et al.*, 1992), and accumulate in the anthers and the pistil. Anthocyanins, flavonols and chalcones also accumulate in anthers. Presumably, the site of action of these flavonoids is in the cytoplasm near the site of synthesis. Although much is known about flavonoids and their involvement

---

[1] Department of Molecular, Cellular, and Developmental Biology, [4]Molecular Biology Institute, 405 Hilgard Avenue, University of California, Los Angeles, California 90095-1606. [2]Department of Plant Pathology, Forbes 204, University of Arizona, Tucson, Arizona 85721. [3]Department of Biology, Loyola Marymount University, Los Angeles, California 90045.

Flavonoids in Cell Function, edited by Béla Buslig and John Manthey.
Kluwer Academic/Plenum Publishers, New York, 2002.

in various plant functions, there is still much to be learned about these compounds. Flavonoids may act in as yet undiscovered ways as internal chemical signals or agents within intact plants.

Flavonoids also have biological activity in animal systems. In animal studies, certain flavonoids have been considered as phytoestrogens and inhibit cancer cell proliferation (Kandaswami *et al.*, 1991; Deschner *et al.*, 1991). Phytoestrogen molecules have been shown to bind to the estrogen receptor in the cell nucleus (Anderson and Garner, 1998; Miksicek, 1993; So *et al.*, 1997). Binding to the estrogen receptor is believed to act primarily by altering the production of specific proteins within the cell. These proteins then bind to regulatory sites on DNA (estrogen response elements) thereby increasing or decreasing the expression of specific genes. Depending on the affinity of the phytoestrogen molecule for the estrogen receptor, the binding of a phytoestrogen to the receptor may result in the partial activation of the receptor (weak agonistic effect) or the displacement of an estrogen molecule, which may reduce receptor activation (antagonistic or anti-estrogenic effect). Non-genomic effects may include any other effects of phytoestrogens to alter the activity of cellular proteins. Phytoestrogens have also been reported to inhibit directly the activity of certain cellular regulatory proteins, including tyrosine kinases (Kuo *et al.*, 1994), which control the activity of other proteins, as well as topoisomerase II, which helps to regulate cell differentiation and the cycle of cell replication. Treatment of a human gastric cancer cell line with genistein almost completely arrested the cell cycle progression at G2-M (Matsukawa *et al.* 1993). Genistein may perturb the process of phosphorylation/dephosphorylation of tyrosine residues of Cdc2 kinase, thereby leading to cell cycle arrest at G2-M. Genistein may also work as a topoisomerase II inhibitor to arrest the cell cycle at G2. In contrast, other flavonoids such as flavone, luteolin, and daidzein, which are structurally very similar to estrogen, arrest the cell cycle at G1 (Bai *et al.*, 1998).

In this work, we measured the concentration and type of flavonoids in antisense-generated mutant plants of alfalfa (*Medicago sativa* L.) and *Arabidopsis thaliana*. These plants are impaired in cell division and differentiation (Woo *et al.*, 1999; Woo, H-H., unpublished results). Flavonoids may be part of the signaling mechanism for cell division and differentiation in these mutant phenotypes.

2. MATERIALS AND METHODS

2.1. Plant Growth and Tissue Collection

Wild-type and transgenic mutant alfalfa and *Arabidopsis* plants were grown in the UCLA greenhouse. Young green leaves and flowers from alfalfa were collected for flavonoid extraction. Roots were collected from 1 month-old stem cuttings of alfalfa plants. *Arabidopsis* plants were grown for 4 weeks before collecting the plant tissues.

2.2. Extraction and Analysis of Flavonoids

Fresh tissues were collected in liquid nitrogen, and the frozen tissues were ground in liquid nitrogen using a mortar and pestle. They were immediately transferred to 100% HPLC-grade methanol and water (9:1 v/v) (1 gram of tissue/20 ml solvent; i.e. 1:20 w/v). The tissues were incubated in the solvent for 30 min at 40°C for flavonoid extraction. After filtration through Whatman No.1 filter paper, the extract was concentrated using a vacuum evaporator at low pressure. All steps were performed quickly because flavonoids can be oxidized rapidly in the air (especially at high pH). For HPLC analysis, crude extracts containing flavonoids were further purified using a reversed phase C-18 drip column. The crude extracts were dissolved in 30% HPLC-grade methanol for the HPLC analysis. Tissue extracts (A_{254}=20) were separated on a Perkin-Elmer high pressure liquid chromatograph equipped with a Nucleosil C-18 column (Phenomenex, 100 X 6.2 mm). Tissue extracts from *Arabidopsis* were analyzed with the following solvents: A = 30% HPLC-grade methanol containing 2.5% glacial acetic acid; B = 100% HPLC-grade methanol containing 2.5% glacial acetic acid. The column was eluted isocratically with solvent A for 5 min and a linear gradient of solvent A to solvent B for 60 min with a flow rate of 1 ml/min. Eluting compounds were monitored at 360 nm. Tissue extracts from alfalfa flowers, leaves, and roots were analyzed with the following solvents: A = 10% HPLC-grade acetonitrile containing 2.5% glacial acetic acid; B = 50% HPLC-grade acetonitrile containing 2.5% glacial acetic acid. The column was eluted isocratically with solvent A for 5 min and a linear gradient of solvent A to solvent B for 60 min with a flow rate of 1 ml/min. Eluting compounds were monitored at 215 nm. All chromatograms were compared with flavonoid standards (e.g., quercetin, kaempferol, and luteolin).

3. RESULTS AND DISCUSSION

3.1. Analysis of Flavonoids in Alfalfa Mutants

Alfalfa mutants were isolated earlier by expressing the antisense RNA of the *PsUGT1*-gene, which encodes UDP-glycosyltransferase (Woo *et al.*, 1999). These mutants were found to exhibit various morphological deviations from wild-type alfalfa such as altered leaf morphology and reduced root growth. The mutant is also sterile for seed production. Further study showed that nodulation by *Sinorhizobium meliloti* is also severely impaired; nodulation in the mutant is about 10% of that of wild-type alfalfa roots (Woo, H.-H, unpublished). All of these altered phenotypes are likely to result from a delay in DNA synthesis, which is brought about the introduction of the antisense construct (Woo *et al.*, 1999). Plants with UGT in the sense orientation do not exhibit the changes in cell cycle time or growth habit.

Sequence analysis showed that the *PsUGT1*-encoded protein has a strong similarity with UDP-glucose:flavonoid-3-*O*-glucosyltransferase (UFGT). This enzyme transfers glucose from UDP-glucose to a flavonoid (Ishikura and Mato, 1993; Gerats, et.al. 1984). An *in vitro* enzyme reaction with recombinant protein of UDP-glycosyltransferase (*PsUGT1*-encoded protein) expressed in *Neurospora crassa* showed that an unknown compound became labeled with UDP-^{14}C-glucuronic acid. Further analysis showed that this unknown compound behaved similarly to flavonoid-derivatives in that it exhibited UV absorption maxima at 270 and 360 nm.

From this observation, we decided to analyze flavonoid contents in the mutant alfalfa lines. Methanol-soluble compounds were extracted from the leaves, flowers, and roots of mutant and wild-type plants for comparison. Separation of the methanol-soluble extracts on a silica TLC plate (F254) showed that compounds with an orange color under long wavelength UV light accumulated in the leaves of the alfalfa mutant (data not shown). The addition of sodium hydroxide to these orange-colored compounds resulted in a batho-chromic shift in all absorption bands, a diagnostic character for flavonoids (Mabry, 1969).

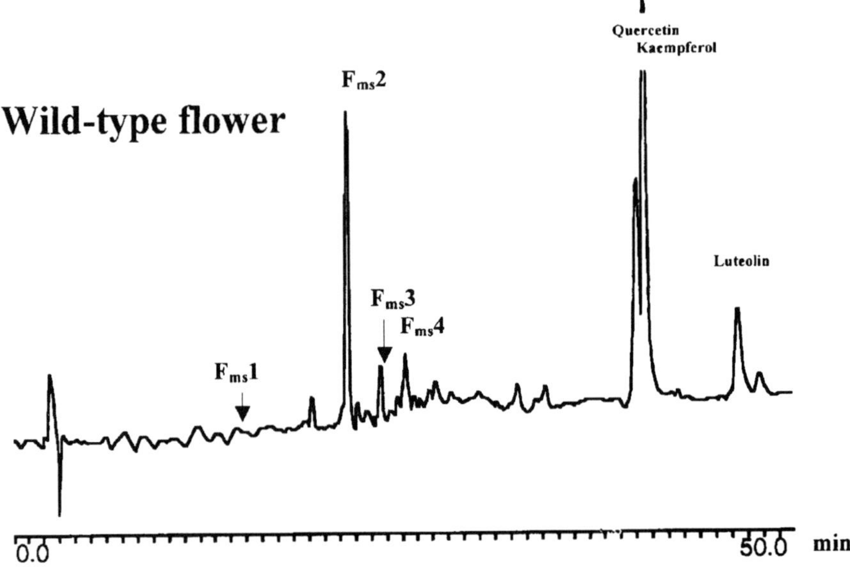

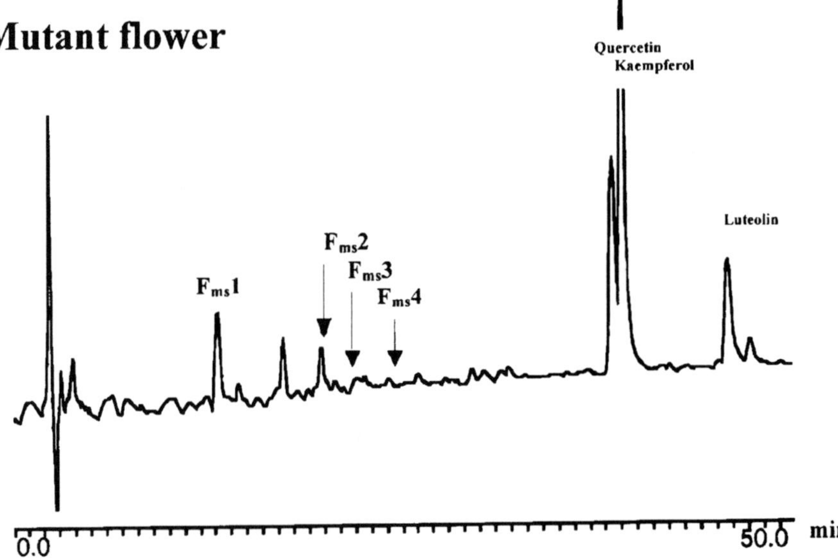

Figure 1. Fractionation of methanol-soluble compounds from alfalfa flower by reverse-phase HPLC. A Nucleosil C-18 reversed-phase HPLC analytical column (Phenomenex) was used. The compounds were eluted at 1 ml/min with a 60 min linear gradient from 87.5:10:2.5 to 47.5:50:2.5, water: acetonitrile: acetic acid (v:v:v). Symbols $F_{ms}1$-$F_{ms}4$ represent UV-absorbing compounds at 215 nm. Compounds $F_{ms}1$ is present at higher levels in mutant flowers than in wild-type flowers. Compounds $F_{ms}2$, $F_{ms}3$, and $F_{ms}4$ are present at higher levels in wild-type flowers than in mutant flowers. Quercetin, kaempferol, and luteolin (2 µg each) were co-injected as an elution standard.

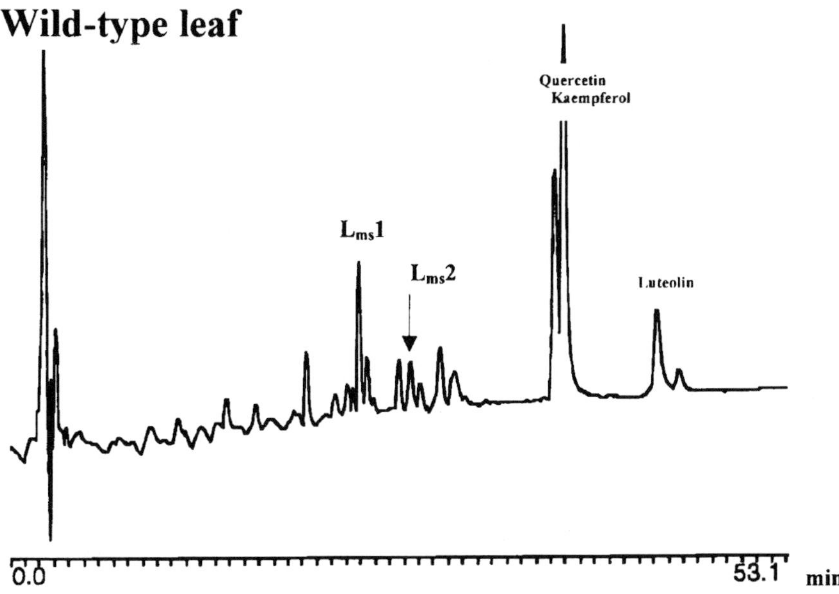

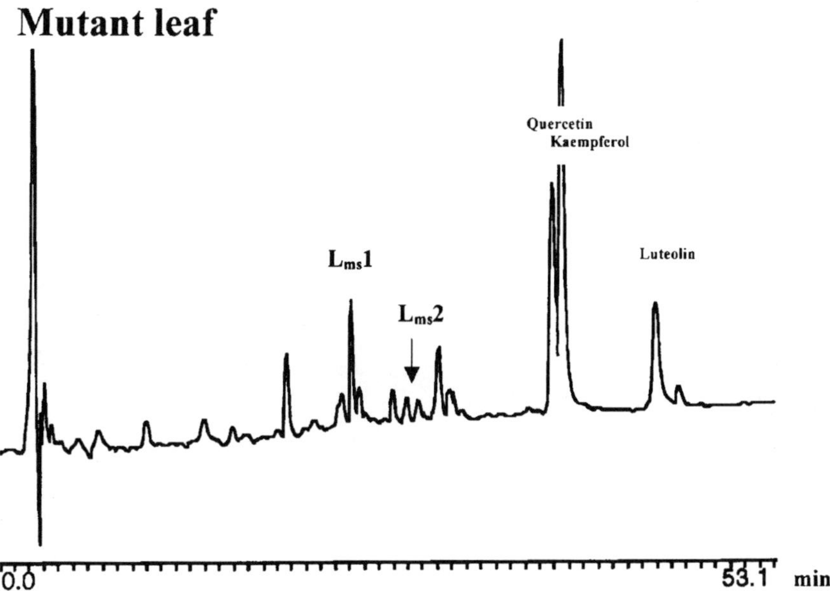

Figure 2. Fractionation of methanol-soluble compounds from alfalfa leaves by reverse-phase HPLC. A Nucleosil C-18 reversed-phase HPLC analytical column (Phenomenex) was used. The compounds were eluted at 1 ml/min with a 60 min linear gradient from 87.5:10:2.5 to 47.5:50:2.5, water: acetonitrile: acetic acid (v:v:v). Symbols $L_{ms}1$ and $L_{ms}2$ represent UV-absorbing compounds at 215 nm. Compounds $L_{ms}1$ and $L_{ms}2$ are present at higher levels in wild-type leaves than in mutant leaves. Quercetin, kaempferol, and luteolin (2 µg each) were co-injected as an elution standard.

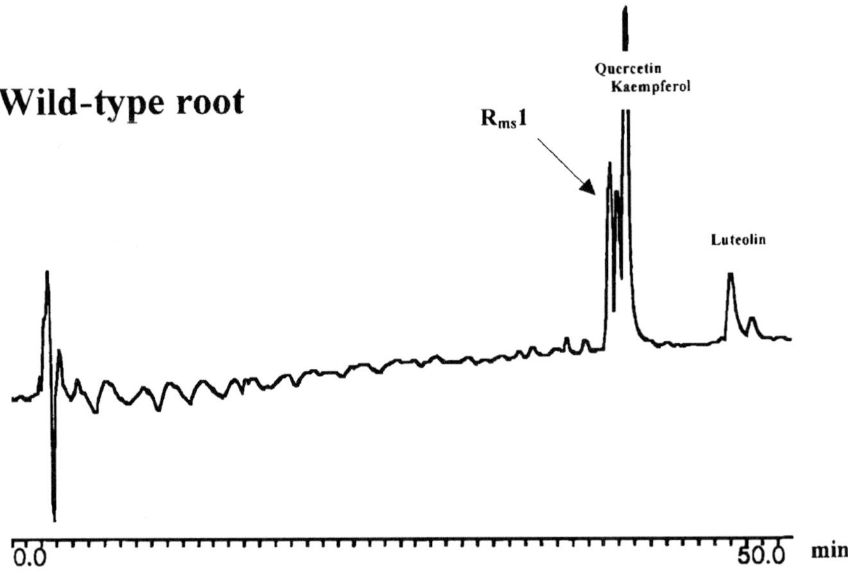

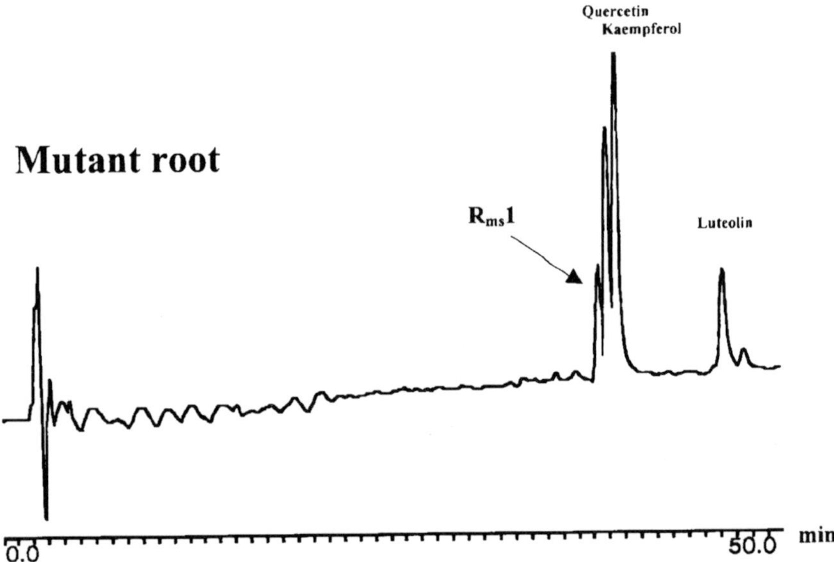

Figure 3. Fractionation of methanol-soluble compounds from alfalfa roots by reverse-phase HPLC. A Nucleosil C-18 reversed-phase HPLC analytical column (Phenomenex) was used. The compounds were eluted at 1 ml/min with a 60 min linear gradient from 87.5:10:2.5 to 47.5:50:2.5, water: acetonitrile: acetic acid (v:v:v). Symbol $R_{ms}1$ represents UV-absorbing compounds at 215 nm. Compound $R_{ms}1$ is present at a higher level in wild-type roots than in mutant roots. Quercetin, kaempferol, and luteolin (2 µg each) were co-injected as an elution standard.

In contrast to the accumulation of orange-colored compounds in the mutant alfalfa leaves, no striking differences in the methanol-soluble compounds from flowers and roots of wild-type and mutant alfalfa were detected using TLC.

Because TLC analysis may not be sensitive enough to detect small changes in amounts or types of flavonoids in mutant tissues, we decided to fractionate the methanol-soluble compounds using HPLC and a reversed phase C-18 column with a UV/ light detector set at 215 nm and 360 nm. HPLC data showed that compound $F_{ms}2$ is present at much higher levels in wild-type flowers than in the mutant flowers (Fig. 1). The compound $F_{ms}2$ may be responsible for the sterility of the mutant flowers. There were also changes both in quality and quantity of few compounds in the mutant leaves (Fig. 2). This change in flavonoid content may be responsible for the altered morphology phenotype in the mutant leaves. In mutant roots, the compound $R_{ms}1$ is present at a lesser concentration than in wild-type roots (Fig. 3). The change in the types of methanol-soluble compounds may be responsible for the delay in DNA synthesis during cell division, reduced nodulation, and sterility for seed production; experiments are in progress to test this hypothesis. HPLC data also showed that there are different compounds in different tissues (i.e. leaves and roots), suggesting that certain flavonoids may have distinct roles in the various tissues of the plant.

3.2. Flavonoids in *Arabidopsis* Mutants

Expression of antisense RNA of UDP-glycosyltransferase in *Arabidopsis* resulted in the accumulation of a red-color (most likely anthocyanins) in the juvenile leaves (Woo, H.-H, unpublished). In contrast, the sense control or wild-type plants remained green in color (Fig. 4). In addition, the mutant plants were shorter and slower to flower than the wild-type

Figure 4. Sense and antisense plants of *Arabidopsis* approximately one month after planting. The antisense plants have a purple color and are slower to develop and flower than the wild-type plants. For a color representation of this figure, see color insert facing page 58.

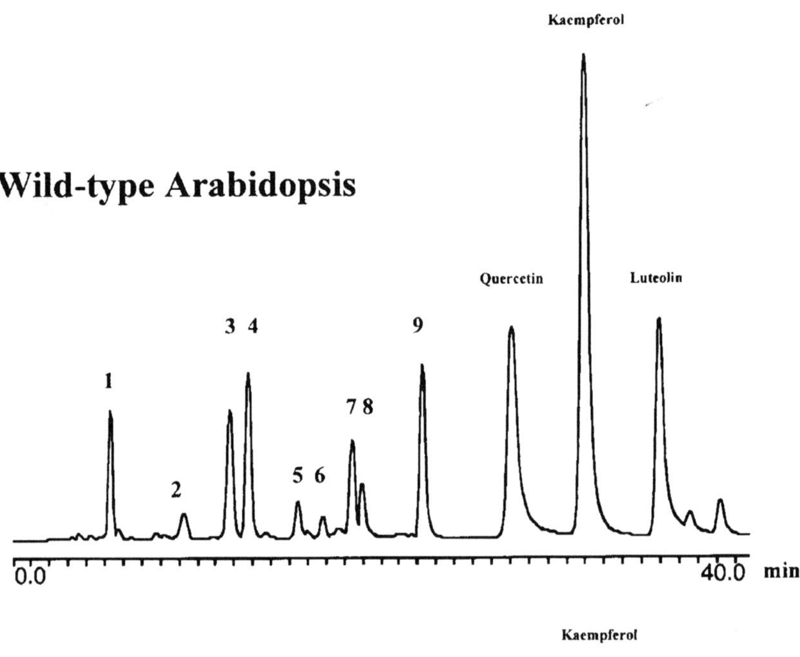

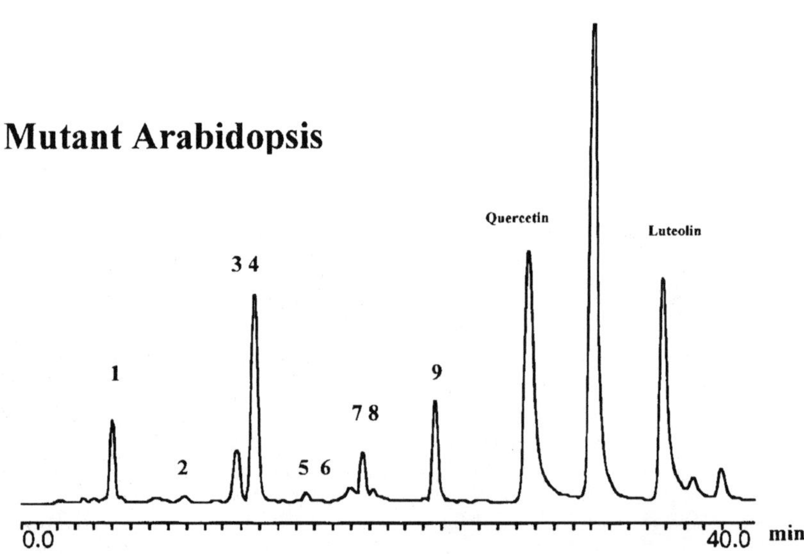

Figure 5. Fractionation of methanol-soluble compounds from *Arabidopsis* by reverse-phase HPLC. A Nucleosil C-18 reversed-phase HPLC analytical column (Phenomenex) was used. The compounds were eluted at 1 ml/min with a 60 min linear gradient from 67.5:30:2.5 to 0:97.5:2.5, water: methanol: acetic acid (v:v:v). Symbols 1-9 represent major 360 nm light-absorbing compounds. Compounds 1-9 except compound 4 accumulate more in wild-type *Arabidopsis* than in mutant *Arabidopsis*. Compound 4 is present at higher levels in mutant *Arabidopsis* than in wild-type *Arabidopsis*. Quercetin, kaempferol, and luteolin (2 µg each) were co-injected as an elution standard.

Figure 4. Sense and antisense plants of *Arabidopsis* approximately one month after planting. The antisense plants have a purple color and are slower to develop and flower than the wild-type plants.

control lines. The mutants were also found to live longer; the life cycle for the mutant *Arabidopsis* is 9-13 weeks compared to 5-7 weeks for wild-type plants. This increased life span of the mutant plants is likely to be caused by the delayed DNA synthesis that is occurring during the cell cycle.

HPLC analysis showed that several compounds except compound 4 were present at higher concentrations in wild-type plants than in mutant plants; compound 4 accumulated more in mutant plants (Fig. 5). Compound 4 may be a substrate of UDP-glycosyltransferase (*PsUGT1*-encoded protein), and compound 3 may be a glycosylated form of compound 4. This result suggests that there are changes in flavonoid levels in the mutant *Arabidopsis*, which may be responsible for the longer life cycle of the antisense plants.

Flavonoids in plants may have effects on gene expression by acting on a putative hormone receptor in the nuclear membrane, or alternatively, they may change the activity of regulatory proteins, such as tyrosine kinase, which are involved in the cell division. Moreover, the altered phenotypes in the mutant plants may be caused by the accumulation of certain flavonoids in various cell types. We are investigating these possibilities.

In conclusion, flavonoids are a diverse group of metabolites that is widespread in higher plants. Flavonoids may have evolved initially to serve a role as internal signal molecules for plant growth and development. Later stages of evolution may have added other functions such as filtering ultraviolet light, serving as chemical signals for symbiosis, and attracting pollinators for flower pigmentation. Future research will focus on the correlation between flavonoid accumulation and effects on the cell cycle.

ACKNOWLEDGMENTS

This work was supported in part by grants to M. C. H. and H.-H. W. from the U.S. Department of Energy, Division of Energy Biosciences, and by grants to M. C. H. and H.-H. W. from the National Science Foundation, Division of Molecular and Cellular Biosciences.

REFERENCES

Anderson, J. J. B., and Garner, S. C., 1998, Phytoestrogens and bone, In *Bailliere's Clinical Endocrinology and Metabolism: Phytoestrogens*, Adlercreutz, H., ed., Bailliere Tindall, London, pp. 543-557.

Bai, F., Matsui, T., Ohtani-Fujita, N., Matsukawa, Y., Ding, Y., and Sakai, T., 1998, Promoter activation and following induction of the p21/WAF1 gene by flavone is involved in G1 phase arrest in A549 lung adeno-carcinoma cells, *FEBS Letters* **437**:61-64.

Baker, M. E., 1992, Evolution of regulation of steroid-mediated intercellular communication in vertebrates: insights from flavonoids, signals that mediate plant-rhizobia symbiosis, *J. Steroid Biochem. Molec. Biol.* **41**:301-308.

Deschner, E. E., Ruperto, J., Wong, G., and Newmark, H. L., 1991, Quercetin and rutin as inhibitors of azoxymethanol-induced colonic neoplasis, *Carcinogenesis* **12**:1193-1196.

Dixon, R. A., and Paiva, N. L., 1995, Stress-induced phenylpropanoid metabolism, *Plant Cell* **7**:1085-1097.

Fisher, R. F., and Long, S. R., 1992, Rhizobium-plant signal exchange, *Nature* **357**:655-660.

Furuya, M., Galston, A. W., and Stowe, B. B., 1962, Isolation from peas of cofactors and inhibitors of indolyl-3-acetic acid oxidase, *Nature* **193**:456-457.

Galston, A. W., 1969, Flavonoids and photomorphogenesis in peas, In *Perspectives in Phytochemistry*, Harborne, J. B., and Swain, T., eds., Academic Press, New York., pp. 193-204.

Gerats, A. G. M., Bussard, J., Coe, E. H., and Larson, R., 1984, Influence of B and Pl on UDPG:flavonoid--3-O-glucosyltransferase in *Zea mays* L., *Biochem. Genet.* **22**:1161-1169.

Harborne, J. B., 1988, Flavonoids in the environment: Structure-activity relationships, *Prog. Clin. Biol. Res.* **280**:17-27.

Ishikura, N., and Mato, M., 1993, Partial purification and some properties of flavonol 3-*O*-glycosyltransferases from seedlings of *Vigna mungo*, with special reference to the formation of kaempferol 3-*O*-galactoside and 3-*O*-glucoside, *Plant Cell Physiol.* **34**:329-335.

Jacobs, M., and Rubery, P. H., 1988, Naturally occurring auxin transport regulators, *Science* **241**:346-349.

Kandaswami, C., Perkins, E., Soloniuk, D. S., Drzewiecki, G., and Middleton, E. Jr., 1991, Antiproliferative effects of citrus flavonoids on a human squamous cell carcinoma *in vitro*, *Cancer Letters* **56**:147-152.

Koes, R. E., and Quattrocchio, F., Mol, J. N. M., 1994, The flavonoid biosynthetic pathway in plants: Function and evolution, *BioEssays* **16**:123-132.

Kuo, M. -L., Lin, J. -K., Huang, T. -S., and Yang, N. -C., 1994, Reversion of the transformed phenotypes of v-H-ras NIH3T3 cells by flavonoids through attenuating the content of phosphotyrosine, *Cancer Letters* **87**:91-97.

Li, J., Ou-Lee, T. -M., Raba, R., Amundson, R., and Last, R. L., 1993. Arabidopsis flavonoid mutants are hypersensitive to UV-B irradiation, *Plant Cell* **5**:171-179.

Mabry, T. J., 1969, The ultraviolet and nuclear magnetic resonance analysis of flavonoids. In *Perspectives in Phytochemistry*, Harborne, J. B. and Swain, T., eds., Academic Press, London and New York., pp. 1-45.

Matsukawa, Y., Marui, N., Sakai, T., Satomi, Y., Yoshida, M., Matsumoto, K., Nishino, H., and Aoike, A., 1993, Genistein arrests cell cycle progression at G2-M, *Cancer Research* **53**:1328-1331.

Miksicek, R. J., 1993, Commonly occurring plant flavonoids have estrogenic activity, *Mol. Pharmacol.* **44**:37-43.

Mo, Y., Nagel, C., and Taylor, L. P., 1992, Biochemical complementation of chalcone synthase mutants defines a role for flavonols in functional pollen, *PNAS* **89**:7213-7217.

So, F. V., Guthrie, N., Chambers, A. F., and Carroll, K. K., 1997, Inhibition of proliferation of estrogen receptor-positive MCF-7 human breast cancer cells by flavonoids in the presence and absence of excess estrogen, *Cancer Letters* **112**:127-133.

van der Meer, I. M., Stam, M. E., van Tunen, A. J., Mol, J. N. M., and Stuitje, A. R., 1992, Antisense inhibition of flavonoid biosynthesis in petunia anthers results in male sterility, *Plant Cell* **4**:253-262.

Vierheilig, H., Bago, B., Albrecht, C., Poulin, M-J., and Piché, Y., 1998, Flavonoids and arbuscular-mycorrhizal fungi, In *Flavonoids in the Living System* (Advances in Experimental Medicine and Biology, vol. 439), Manthey, J. A. and Buslig, B. S., eds., Plenum Press, New York and London., pp. 9-33.

Weiss, M. R., 1991, Floral color changes as cues for pollinators, *Nature* **354**:227-229.

Woo, H. -H., Orbach, M., Hirsch, A. M., and Hawes, M. C., 1999, Meristem-localized inducible expression of a UDP-glycosyltransferase gene is essential for growth and development in pea and alfalfa, *Plant Cell* **11**:2303-2316.

MODERN ANALYTICAL TECHNIQUES FOR FLAVONOID DETERMINATION

Mark A. Berhow[1]

1. INTRODUCTION

The biological functionalities of plant natural products, both in plants and in the animals that use and consume them, is fueling new interest in phytochemical research (Cordell, 1995). A key component of this research is the ability to accurately identify and quantitate specific phytochemicals. Today analytical equipment is available that can rapidly separate, unequivocally identify, and accurately quantify phytochemicals from plant materials literally in a matter of minutes. The development of low-cost, high-powered computer systems allowed for the creation of computer-driven bench top chromatography instrumentation. These systems are able to perform complicated electronic functions, such as control of gas flow and liquid pumps, spectral detection, including optical and mass spectrometry and to accumulate, quantitate, and process large amounts of data. Previously, there may have only been one or two of these types of instruments at an institution. Now they are available at prices in the $20,000 to $200,000 range, which makes them affordable to individual researchers. These developments have made complex plant product analysis extremely practical and affordable.

To perform natural product analysis requires three basic steps: an efficient method of extracting the phytochemicals of interest from the raw plant material, a way to separate the various phytochemicals in the extracts, and a method of identifying and quantitating the phytochemicals. For this, pure standards are essential. Despite the thousands of phytochemicals that have been identified and reported in the literature, commercially available phytochemical standards are still rarely available and usually quite expensive. A key component of phytochemical analytical methodology is still, therefore, the isolation of pure standards. Fortunately, a great deal of progress has been made in improving chromato-

---

[1] U. S. Department of Agriculture, Agricultural Research Service, National Center for Agricultural Utilization Research, 1815 North University Street, Peoria, IL, 61604 , USA. E-mail: berhowma@mail.ncaur.usda.gov

Flavonoids in Cell Function, edited by Béla Buslig and John Manthey.
Kluwer Academic/Plenum Publishers, New York, 2002.

graphic matrices over the past 10 years, so the purification of chemicals is more efficient and larger segments of the purification process can be automated to yield milligram quantities of purified compounds in days or even hours.

Flavonoids are synthesized from phenylalanine via a biosynthetic route in plants termed the general phenylpropanoid pathway (Heller and Forkmann, 1994). Secondary metabolites derived from this pathway, in addition to flavonoids, include tannins, phenols, benzoic acids, stilbenes, cinnamate esters, and coumarins. Flavonoids have generally been classified into 12 different subclasses by the state of oxidation and the substitution pattern at the C-2-C-3 unit. These include flavanones, flavones, flavonols, chalcones, dihydro-chalcones, anthocyanidins, aurones, flavanols, dihydroflavonols, proanthocyanidins (flavan-3,4-diols), isoflavones, and neoflavones. More than 10,000 flavonoids have been identified from natural sources and more continue to be identified at a rate of more than 10 per month (Dakora, 1995). Different plant species often vary widely in flavonoid type and content (Seigler, 1981).

The biosynthesis of flavonoids is constitutive in most plants, controlled internally during normal growth and development, such as new vegetative leaf growth and repro-ductive organ development (Heller and Forkmann, 1994). Plants will produce and accu-mulate different flavonoids depending on several factors, such as plant growth stage, reproductive stage, the particular plant tissue involved, and the type of environmental stress or pathogenic attack involved (Dixon, 1999). These compounds play important roles in the interaction of the plant with its environment (Berhow and Vaughn, 1999).

Most flavonoids are accumulated and stored in appreciable quantities in the plant cells as *O*- and *C*-glycosides. Flavonoid biosynthesis may also be induced by exogenous stimuli, such as changes in light and temperature (Hahlbrock and Scheel, 1989). Such induced compounds accumulate specifically in tissues around the site of the stimulus or injury, often at very high concentrations (Dixon and Paiva, 1995). In soybean plants the three main storage flavonoids are the isoflavones genistein, daidzein, and glycitein (see Fig. 1 for struc-tural details). They accumulate as glucoside and malonyl-glucoside conjugates in the seeds (Kudou et al., 1991; Barnes et al., 1994; Wang and Murphy, 1994, 1996; Tsukamoto et al., 1995; Garcia et al., 1997; Song et al., 1998; Wang et al., 1998; Wang et al., 2000). These may be modified by cooking and processing to the acetyl-glucosides and even the aglycones (Coward et al., 1998). In soy, one of more of the storage isoflavonoids may be converted to the biologically active stress response compounds such as medicarpin and biochanin A at the point of stress (Dixon and Paiva, 1995).

Of current interest is the analysis of flavonoids in animal and human fluids and tissues. Ingestion of foods containing flavonoids result in the absorption, circulation, metabolism, and eventual excretion of these compounds (Barnes et al., 1999; Shelnutt et al., 2000). The dietary flavonoids may have profound effects on the health of the animal that consumes them, through their antioxidant and estrogenic activities, and a wide spectrum of anti-microbial and pharmacological properties (Messina, 1999). They may have an effect on some of the chronic diseases associated with aging, such as arthritis, coronary heart disease, and cancer.

Several good reference guides have been published on the extraction, separation, and analysis of plant natural products. These include a series on the flavonoids edited by Harborne and others (1975, 1982, 1988, 1994), and reviews on the analysis of plant natural products by Colegate and Molyneux (1993), and by Mann et al. (1994).

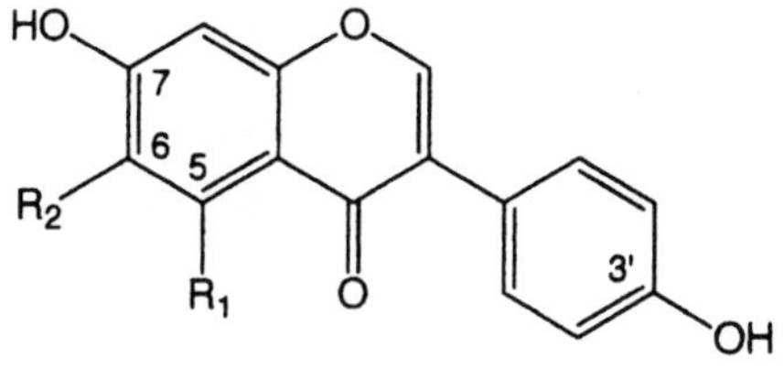

daidzein (R1 = H, R2 = H)
glycitein (R1 = H, R2 =OMe)
genistein (R1 = OH, R2 = H)

In plant tissues:

Daidzin (7-O-glucosyl-daidzein)
Glycitin (7-O-glucosyl-glycitein)
Genistin (7-O-glucosyl-genistein)

6"-O-malonyl-7-O-glucosyl-daidzein
6"-O-malonyl-7-O-glucosyl-glycitein
6"-O-malonyl-7-O-glucosyl-genistein

Additional forms found in processed food products:

6"-O-acetyl-7-O-glucosyl-daidzein
6"-O-acetyl-7-O-glucosyl-glycitein
6"-O-acetyl-7-O-glucosyl-genistein

Figure 1. Structures of isoflavones found in soybean seeds and processed food products.

2. PREPARATION OF SAMPLES FOR ANALYSIS

There are a wide variety of methods to prepare samples for analysis, many are needlessly complex (Hasegawa et al., 1995). Plant tissues can be extracted fresh, but more often the tissues are dried or freeze-dried and ground to a fine powder to increase the efficiency of extraction. Classical bulk preparation methods include the use of Soxhlet extractors, or small samples can be extracted by simply mixing the sample and solvent in a test tube and filtering the resulting liquids. Newer methods include the use of contained high pressure and high temperature systems and supercritical fluid extraction (SCF). Tissues that contain high levels of oils, such as seeds, can be initially extracted with nonpolar solvents such as hexane to remove them. The flavonoids are often found in plant tissues in glycosylated forms. These can be extracted with water or polar organic solvents such as acetone, ethyl acetate, acetonitrile, methanol, ethanol, etc., usually heated to increase the efficiency of extraction. The aglycones can also be extracted in this manner. As some of the flavonoid glycosides are not very soluble in many solvents, dimethylsulfoxide or dimethylformamide can be used in the extraction cocktail. Animal tissues and fluids can be extracted in a similar manner.

Contaminating carbohydrates, proteins, ionic compounds, and nonvolatile solvents can be removed by passing aqueous samples through various solid phase extraction (SPE) columns that are available from many different commercial companies. In general, we use reverse phase C18 mini columns. Under the aqueous conditions, the flavonoids are retained on the column, while many, if not most, of the water soluble contaminants are not. Organic solvent extracts such as methanol, DMSO, and acetonitrile can be diluted about 10 fold with water and passed through a small SPE column with a syringe. The flavonoids are retained on the column and, after rinsing with water, can be eluted with an appropriate volatile solvent such as methanol. Care must be taken to insure the retention of all the flavonoids, especially the conjugated forms that carry a negative charge. This is usually done by the addition of a counter ion, such as an acid. Once the samples are in a volatile organic phase, they can be concentrated for spotting on TLC plates or filtered for injection onto HPLC columns.

If necessary, the glycosidic or other ester residues can be removed by enzymatic hydrolysis using a variety of specific or non-specific glycoside hydrolyzing enzymes that are now commercially available, or by acid hydrolysis, refluxing in 2 M hydrochloric or sulfuric acid. The flavonoid aglycones can be partitioned into a water immiscible solvent such as ethyl acetate or diethyl ether.

3. COLORIMETRIC AND SPECTROPHOTOMETRIC DETECTION METHODS

The most convenient method to estimate the concentration of pure flavonoids is by their UV absorbance. The spectral characteristics of most of the classes of flavonoids have been published (Mabry et al., 1970). A standard curve using a pure standard at an appropriate wavelength will give an estimate of the total UV absorbing material in an extract (or a rough estimate of the phenolics of interest at that wavelength). In order to quantitate a particular flavonoid, interfering compounds must be removed and the individual flavonoids separated. The usual methods use either column chromatography or thin-layer chromatography (TLC) as will be discussed further in sections below.

UV absorption can also be used to rapidly estimate flavonoid content in samples, e. g., the measurement of flavonoids in citrus juices (Hendrickson et al., 1958). The sample is diluted/extracted with 2-propanol and filtered to obtain a solution that can be measured in a spectrophotometer. The difference between absorbances at 295 and 290 nm is proportional to the flavonoid content. Other methods can be more specific and rely on reagents to shift the absorbance spectra, such as the Davis Method for flavanones employing alkaline diethyleneglycol reagent (Davis, 1947), measuring the pink color flavonoids develop in the presence of magnesium and hydrochloric acid (Kwierty and Braverman, 1959), the red color developed by flavonoids when treated with 1-nitroso-2-naphthol and nitric acid (Gengross and Renda, 1966), and using an enzyme assisted colorimetric measurement with 2,2-azino-bis-(3-ethyl-benzthiazoline-6-sulfonic acid) (Arnao et al., 1990).

4. THIN-LAYER CHROMATOGRAPHY (TLC) SEPARATION AND DETECTION

There are many TLC protocols that can be used to separate flavonoids that have been extracted from plant tissues. Methods include the use of silica, reverse phase or polyamide plates in a wide variety of solvent systems. Many of these solvent systems are published in a chapter on phenolic compounds in Kirchner's excellent book *Thin Layer Chromatography* (1978). The flavonoid spots on the developed plates can be visualized in a variety of manners, the most common method being the direct visualization under UV light using plates that have an acid-stable fluorescent indicator added for visualization at 254 nm. With the use of standards, visualized spots can then be scraped off, extracted from the silica with organic solvent, and quantified by UV absorption. A visualization method for flavanones utilizes a spray of 2% sodium borohydride freshly prepared in methanol followed by fumigation with hydrogen chloride gas (Horowitz, 1957). Other visualization methods are outlined in Kirchner's book.

TLC methods usually vary from one another mainly in the extraction procedures. We have found that one of the best solvent systems for developing silica plates with flavonoid glycosides is ethyl acetate-methylethylketone-formic acid-water (5:3:1:1) and that for developing polyamide plates is nitromethane-methanol (2:1), though there are many others (Hasegawa et al., 1995).

5. HIGH-PERFORMANCE LIQUID CHROMATOGRAPHY (HPLC) SEPARATION

5.1. HPLC with Photometric Detection

HPLC is currently the method of choice to accurately determine both the composition and the absolute concentrations of the flavonoids in a given sample. Since HPLC was developed in the 1960s, its use has been expanded and improved to evaluate flavonoids in extracts from all parts of the plant. All HPLC systems use one or more pumps, possibly a gradient device, an injector, and a detection system, such as a UV/VIS spectrophotometer or a mass spectrometer. Today HPLC systems are fully computer controlled and automated, with automatic injection systems, column and solvent temperature control, photodiode array detectors (PDA, which collect both UV-VIS traces and a full spectra of each peak detected), programmable fraction collectors and much more. The packings used in today's columns are much more stable to changes in pH, to the various organic solvents, and generally more rugged than those of even ten years ago.

The chromatography literature abounds in HPLC methods, using different isocratic, binary, ternary, and quaternary gradients, on both reverse and normal phase columns. Some of these methods were reviewed as they pertain to citrus flavonoids (Hasegawa et al., 1995). Most of the HPLC analyses are run using C18 (ODS) reverse phase columns with either isocratic or gradient elution. Isocratic systems include simple acetonitrile-water systems which is now usually enhanced by the addition of a small amount of acid. Isocratic water-methanol-acetonitrile-acetic acid, methanol-0.01 M phosphoric acid, water-acetonitrile-tetrahydrofuran- acetic acid have also been used to separate flavonoids. Identification and quantification is usually achieved by monitoring the absorbance between 280-290 nm and comparing to known standards, but retention times of flavonoids in complex mixtures can vary considerably from those of separately run standards. The resolution limit is on the order of 1 microgram, but samples can be concentrated to overcome that limit. One of the more frequently cited HPLC methods for the analysis of soy isoflavones is that of Wang and Murphy (1994). This method utilizes a C18 reverse phase column and the column is developed with a programmed gradient of 15% acetonitrile (ACN) in water (with 0.1%

glacial acetic acid) to 35% ACN over 50 minutes at a flow rate of 1 mL/min.. The detection limits are just over 100 ng/mL.

Various reverse phase packing materials can be used, especially the high purity, consistent pore sized, super-deactivated silica that are available today. There are a variety of particle sizes, pore sizes, and types (C18, C8, phenyl, cyano, amino, and more). We have been using a RP C18, 3 micron, ODS 3 column which gives excellent peak separation and reproducibility as well as being rugged and long lived. Our column development gradient is 20%, methanol in 0.01 M phosphoric acid developed to 100% methanol over 50 minutes. This gives an excellent separation of the isoflavonoids found in soy extracts (Fig. 2). The

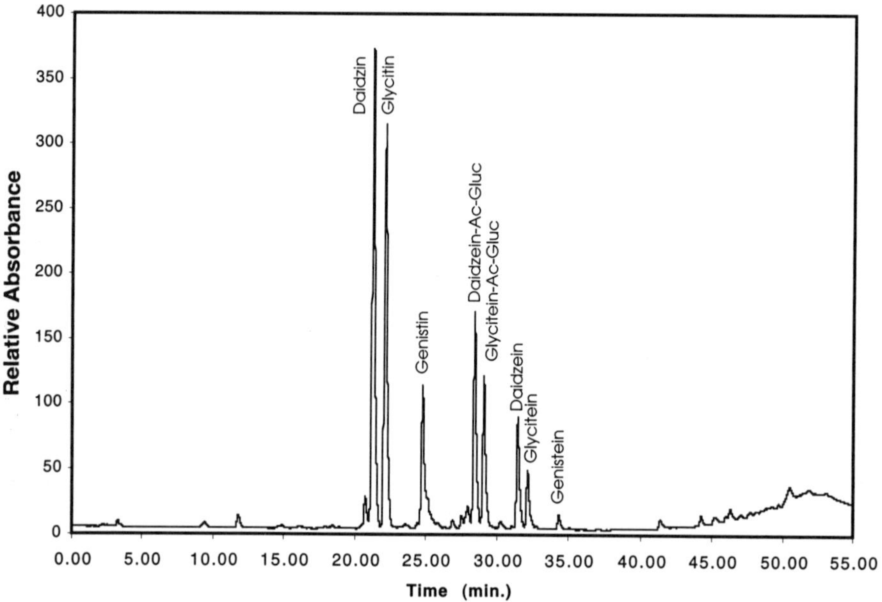

Figure 2. Trace of HPLC RP-C18 analysis of isoflavonoids in a commercial isoflavone concentrate prepared from soybean endosperm. Chromatogram was developed under conditions described in the text. The effluent was monitored at 285 nm with a PDA detector.

advent of photodiode array detector systems increases the amount of information obtained in a run. Full spectra can be obtained at several points during the elution of a peak. This can be used to identify the different classes of flavonoids. The spectra for daidzin, glycitin, and genistin are shown in Fig. 3. Computer software can be used to identify and assign peaks based on retention times and spectra. The spectra can also used to determine the degree of purity in a peak as well. In general, the spectra of the aglycone, the glucoside, and the acetyl glucoside matches very closely (Fig. 3). This means that pure aglycone standards can be used to quantitate all the glucoside forms on a molar basis.

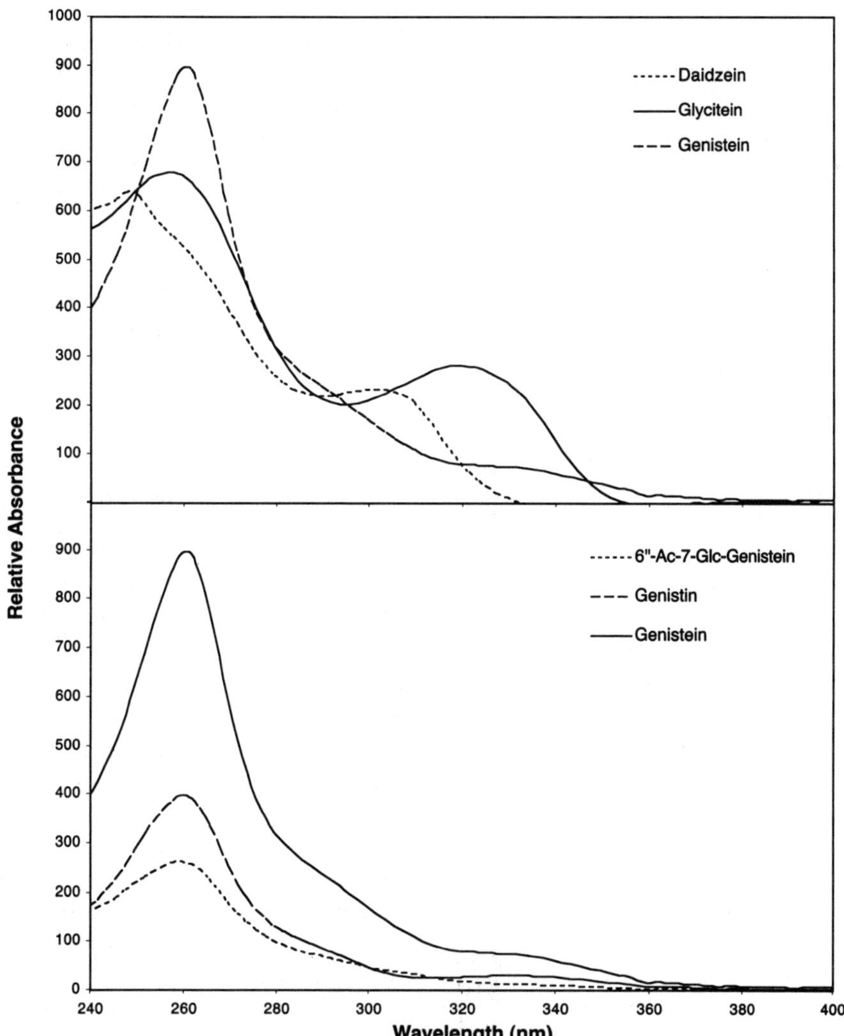

Figure 3. Characteristic UV spectra of selected soy isoflavones. Upper panel shows the spectra of the isoflavone aglycones. The lower panel shows the spectra of genistein aglycone, genistein glucoside (genistin) and genistein acetyl-glucoside.

5.2. HPLC with Mass Spectrometric Detection (LC-MS)

LC-MS is rapidly becoming the method of choice for natural product analysis. Mass spectrometry (MS), along with nuclear magnetic resonance (NMR) spectrometry, are the methods applied to the structure determination of phytochemicals (Colegate and Molyneux, 1993). The advantages of LC-MS over other MS methods include the lack of chemical derivatization, the increased sensitivity and that the specificity of the MS, which is superior to that of UV-VIS spectrometry. During the course of MS analysis, chemicals are ionized and the resulting charged species are separated according to their mass (m) to charge (z) ratio (m/z). This allows for considerable flexibility in detection, quantitation, identification and structural determination of compounds. The application of MS to the analysis of plant natural products has increased with the development of the thermospray atmospheric pressure ionization (so called soft ionization) techniques during 1970-1990 (Careri et al., 1998; Jungbuth and Ternes, 2000; Stobiecki, 2000).

The two main atmospheric ionization techniques that emerged (and now found on most LC-MS systems today) are electrospray ionization (ESI) (Zhou and Hamberger, 1996) and atmospheric pressure chemical ionization (APCI) (Wolfender et al., 1995). These techniques produce singly or multiply charged gaseous ions directly from an aqueous or aqueous/organic solvent system. In the case of APCI, the solvent is vaporized off, putting the chemical particles in gaseous phase before the addition of the charge, while in ESI the charge is introduced in the liquid state as the eluate emerges in a spray from the column, and dry gas and heat cause the solvent to evaporate. The result is an introduction of a charge to the chemical species which is then drawn into the mass analyzer by a vacuum through a series of focusing lenses. The detector can be set up to detect either positively or negatively charged ion species.

The success of the technique strongly depends on the LC mobile phase composition and flow rate. Typically flow rates of 0.3 mL/min or lower are used, along with narrow bore (3 mm) or microbore (1 mm) columns. The development of the better solid phases allows for the use of narrower columns. The mobile phase usually has a source of easily ionized volatile compounds, such as ammonium acetate, ammonium formate, acetic acid, and trifluoroacetic acid, from which a charge is transferred to the phytochemical. This generates the charged moieties such as $[M+H]^+$. Other cation molecules, such as NH_4^+, Na^+, or K^+ ions are also registered with adducts of components of the mobile phase. The soft ionization provides minimal fragmentation energy, but often, especially in ESI some of the glycosidic moieties will fragment off the flavonoid compound and yield information on the composition of the glycoside moiety. Coupling the APCI and ESI interfaces to triple quadruple or ion trap mass spectrometers allows for the generation and analysis of daughter fragments in MS-MS and MS^n analysis. Both APCI and ESI techniques have been applied to the analysis of the isoflavones extracted from plant material and from animal sera and excretions, using the methodology described above. The use of the negative mode allows for more sensitivity and can be used to quantitate the compounds, with minimal or even no separation. The use of selective ion monitoring in today's software provides this powerful quantitation alternative to UV absorption. A word of caution, however, the response curve for these compounds in the MS is not completely linear.

The application of LC-MS has been extensively applied to the analysis of flavonoids in soybeans and related legumes (Barnes et al., 1994, Barnes et al., 1998; Andlauer et al., 1999; Barnes et al., 1999; Careri et al., 1999; Stobiecki et al., 1999; Lin et al., 2000). We have performed flavonoid analysis with an ion trap MS coupled to a low-flow HPLC and a PDA detector. The entire system operates under the control of a software package. We are using an Inertsil reverse phase C18 column (3μ, ODS 3, 3 x 150 mm), though any other high purity, fully endcapped packing material will give good resolution. The column is developed with a binary gradient starting at 20% methanol and 80% water, both solvents

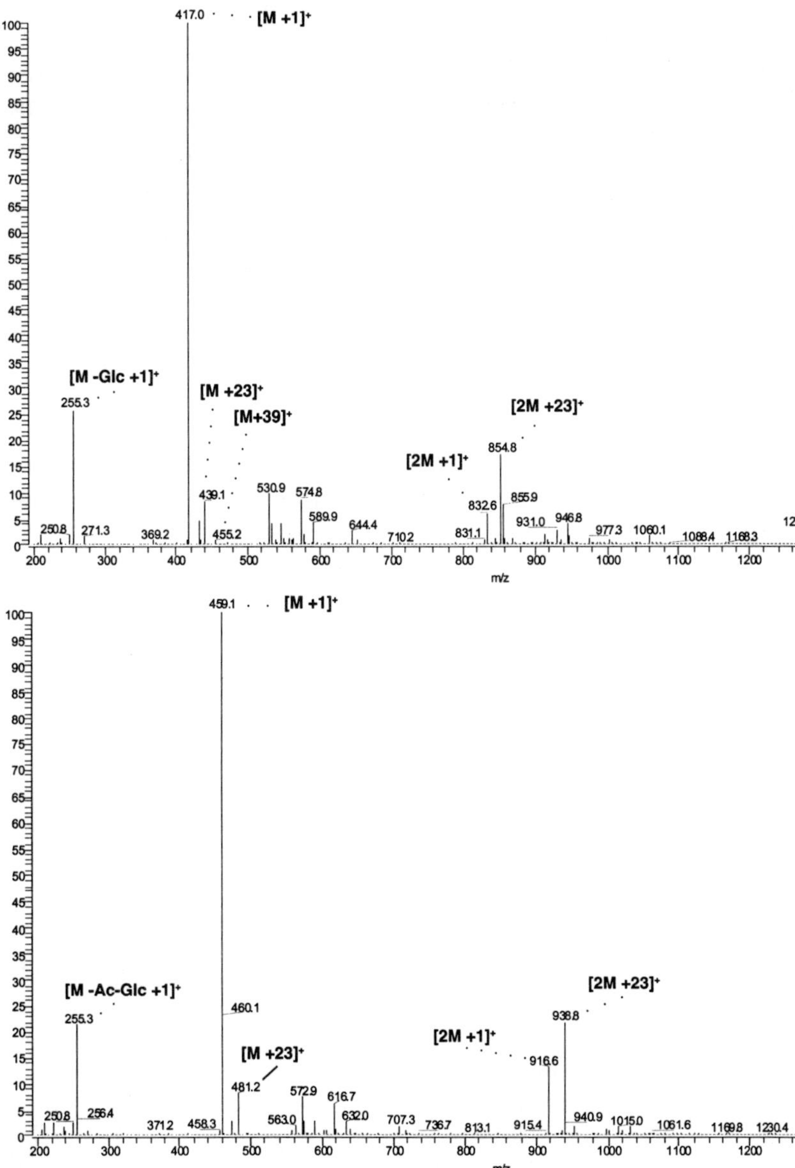

Figure 4. Positive ion ESI-MS of daidzin (upper panel) and daidzein acetyl-glucoside (lower panel) showing characteristic mass ions present.

contain 0.25% acetic acid, with a linear gradient to 100 % methanol over 55 minutes at a flow rate of 0.3 mL/min. Care must be taken to avoid plastic bottles, transfer pipettes, etc., to avoid contamination with plasticisers which will show up in the MS. The MS software allows for a calibration procedure which will optimize for the detection of the flavonoids of interest. If one is looking for a known, this calibration can be done using either a standard or the known mass ion within the sample. Another key point is not to inject too much sample. Under normal MS conditions (that is with little or no collision energy introduced), key diagnostic ions can be detected for the molecules of interest. In the positive mode these include $[M+1]^+$ and $[M\text{-glucose}+1]^+$ for the soy isoflavones. Depending on sample preparation and the degree of contamination in the HPLC system, both sodium $[M+23]^+$ and potassium $[M+39]^+$ adducts may also been seen. Spectra for daidzin and 6"-*O*-acetyl-7-*O*-glucosyl-daidzein are shown in Fig. 4. Using the ESI mode, we have also detected dimers $[2M+1]^+$ and $[2M+23]^+$. These have been reported for flavonoids found in citrus (Swantsitang et al., 2000). The LC-ESI-MS analysis of the flavanone naringin shows the presence of dimers $[2M+1]^+$ and $[2M+23]^+$, plus a dimer formed with contaminating iron $[2M+55]^+$, as shown in Fig. 5. Flavanones will form stable complexes with certain metals such as aluminum which are detectable in the MS (Deng and Van Berkel, 1998). The complexes formed in the ESI inter-face are due to combination of charge and reduction of the liquid spray globules forming complexes with whatever cation is present. The dimers and trimers formed are an additional diagnostic that can be used to confirm species identification.

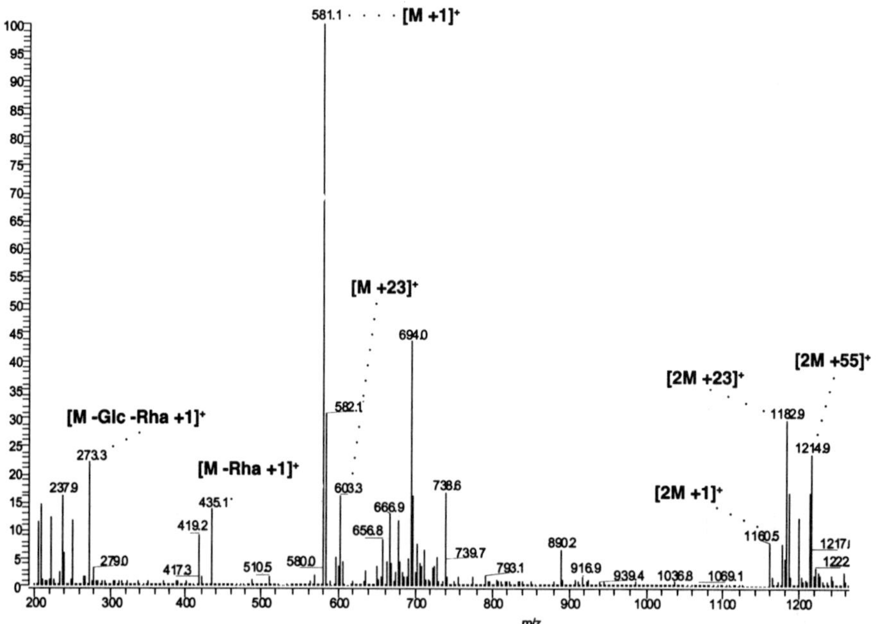

Figure 5. Positive ion ESI-MS of the flavanone naringin (naringenin-7-*O*-neohesperidoside).

In general, LC-MS, when coupled with a spectrophotometric detector, is an excellent technique for the rapid separation, identification, and quantitation of flavonoids in an extract.

6. OTHER SEPARATION AND DETECTION METHODS

6.1. Radioimmunoassay (RIA) and Enzyme-Linked Immunoassay (ELISA or EIA) Quantitation

Jourdan et al. (1982) reported the development of a RIA to detect the presence of the bitter flavanone neohesperidosides. The assay uses an immobilized antibody raised in rabbit against a naringin-bovine serum albumin conjugate, which is specific for the flavonoid neohesperidosides. Samples are added to wells containing the immobilized protein bound to the antibody. The amount of bound material is determined by saturating the remaining antibodies with either ^{3}H-labeled or ^{125}I-labeled naringin. This method can be used to detect naringin down to 2 nanograms. The assay combines both high sensitivity, and rapid, easily performed assays. RIAs have been used to analyze naringin levels in *Citrus* tissue (Jourdan et al., 1985a,b). A nonradioactive form of this assay, the enzyme-linked immunoassay which utilizes naringin linked to alkaline phosphatase was also developed by Jordan et al. (1984) and used by Matsumoto and Okudai (1991).

While immunoassays are extremely sensitive and fast, they do have a number of interfering agents and cannot accurately discriminate between the different flavanone glycosides. In addition, the components for the assay, such as specific antibodies for specific flavonoids, the radioactively labeled flavonoids or the alkaline phosphatase linked flavonoids, are not commonly available.

6.2. Capillary Electrophoresis Separation and Detection

Capillary electrophoresis (CE) is becoming increasingly recognized as an important analytical separation technique. It has the advantage of being fairly rapid (runs of about 10 minutes) and simple, similar to the HPLC methodology. One of the first protocols was published for the separation of flavonol-3-glycosides from *Ginkgo biloba* (Pietta et al., 1991). Seitz et al. (1992) published a CE protocol for the separation of flavonoids from *Sambuci flos*. This technique has been applied to the analysis of isoflavones and other flavonoids (Shihabi et al., 1994; Aussenac et al., 1998; Runkel et al., 1998; Cancalon, 1999; Voirin et al., 2000). As sample preparation and chromatography conditions are fairly standardized, these techniques are easily modified for the analysis of flavonoids from any plant species.

6.3. Gas Chromatography-Mass Spectroscopy Separation and Detection

Because the flavonoid glycosides are nonvolatile, GC has not been routinely been used for flavonoid analysis. Creaser et al. (1992) have developed a method to analyze the hydrolyzed flavanone content by GC-MS. The method involves solvent extraction of the tissues, hydrolysis of the flavonoids to their respective aglycones, trimethylsilation, and analysis by GC-MS. While the method will accurately identify and quantify the flavonoid aglycones present in the sample, it cannot determine the relative amounts of the neohesperidosides and rutinosides and would therefore only be able to give an approximation of the amount of bitter flavonoids in a sample. GC-MS analysis has been applied to flavonoids from several plant sources in recent years (Bocchini et at., 1998; Kohen et al., 1998; Watson and

Pitt, 1998; Novotny et al., 1999). It has also been applied to the analysis of flavonoid aglycones in blood plasma samples (Luthria et al., 1998; Watson and Pitt, 1998; Heinonen et al., 1999).

7. SOURCES OF FLAVONOID STANDARDS

Thousands of flavonoids have been isolated and identified from the tissues of plant species. Many of these are not commercially available. However, all the large chemical companies, such as Sigma, Aldrich, Fluka, etc., sell a few of the more common flavonoids such as naringin, hesperidin, naringenin, hesperetin, rutin, quercetin and a few others. There are at least three chemical companies that sell a much wider selection of flavonoid chromatographic standards - Extrasynthese SA (BP 62, 21 Lyon Nord 69730 Gernay, France) and Carl Roth GmbH + CoKG Chemiische Fabrick (Schoemperlenstrasse 1-5, Postfach 211162, D 7500 Karlsruhe 21 Germany), distributed in North America by Atomergic Chemetals Corp., (91 Carolyn Blvd., Farmingdale, NY 11735, USA), and Indofine Chemical Company (P. O. Box 473, Somerville, NJ 08876 USA). In the case of the soy isoflavones, genistein, daidzein, daidzin, and genistin are commercially available, but only in mg quantities. A company called LC Laboratories (Woburn, MA 01801; www.LCLabs.com) is currently offering mg quantities of purified standards of all 12 forms of isoflavones found in soybeans.

8. PREPARATION OF FLAVONOID STANDARDS

Unequivocal identification and quantitation of flavonoids in extracts from plants and animals require the comparison with pure standards. Unless one is lucky and can get a set of standards from a commercial company or a fellow researcher, one will be faced with the need to purify phytochemical standards. The recent advances in equipment and chromatography matrices now make the isolation of these compounds an easier task.

One of least expensive methods for isolating pure compounds is the use of preparative TLC. Equipment and supply purchases are minimal in comparison to purchasing and maintaining a preparative HPLC system. Material can be extracted with aqueous and/or organic solvents, then crude separations can be done with solvent partitioning, SPE, and open column chromatography as outlined in the above sections. Concentrated fractions can then be spotted on TLC plates, developed, scraped and extracted from the silica as noted in Kirchner's book on TLC (1978). This method can generate mg quantities sufficient to use as standards in HPLC-PDA and LC-MS.

Larger scale isolations can be done with preparative HPLC. Using a commercially prepared isoflavonoid concentrate, we were able to separate the isoflavones from contaminating carbohydrates and saponins on open C18 reverse phase column developed with increasing amounts of organic solvent. The fractions were monitored by HPLC and TLC to determine the best separation of isoflavones from the saponins. The dried isoflavone fractions were resuspended in a small amount of DMSO:MeOH, diluted to 25% organic solvent with water, and loaded on a preparative HPLC system equipped with a Waters radial compression PrepLC assembly using a Waters Bondapak C18 column cartridge (15-20 um, 125A, 47 x 300 mm). The column was developed isocratically with 25% methanol in water and 1% acetic acid at a flow rate of 50 mL/min. The isoflavonoids eluted with sufficient separation to allow for the isolation of daidzin, glycitin, genistin, daidzein-6"-*O*-acetyl-7-*O*-glucoside, glycitein-6"-*O*-acetyl-7-*O*-glucoside, and genistein-6"-*O*-acetyl-7-*O*-glucoside. The methanol was removed from the fractions with a rotary evaporator, and passed through a C18 SPE, washed with water to remove the acid, and eluted with MeOH. Water was added to the eluate and the purified fractions were allowed to evaporate in a hood. If sufficient compound is present, the pure compound will crystallize out of the solution. Care must be taken to avoid cross contamination in the case of glycitin and daidzin as these two compounds will co-crystallize. Aglycone standards can be made by hydrolyzing the glucosides by refluxing with 2M HCl in MeOH for 24 hours, and removing the acid and carbohydrates through a C18 SPE. Wang and Murphy (1994) outlined a method for the purification of the malonyl conjugates, which are extremely labile.

9. SUMMARY

We stand at the dawn of the next great period of natural products research. We now have tools to begin unraveling the chemical and physiological mechanisms by which these phytochemicals affect the function of living cells, both in plants and animals. The advances in analytical research will allow us to better use these compounds in the control of diseases in both plants and animals. This chapter used the analysis of soy isoflavones to illustrate the approaches to natural product research, but these principles can be applied to nearly any of the phenolic compounds found in plants.

10. REFERENCES

Andlauer, W., Martena, M. J., and Fürst, P., 1999, Determination of selected phytochemicals by reverse-phase high-performance liquid chromatography combined with ultraviolet and mass spectrometric detection, *J. Chromatog. A* 849:341-348.

Arnao, M. B., Casas, J. L., Del Rio, J. A., Acosta, M., and Garcia-Canovas, F., 1990, An enzymatic colorimetric method for measuring naringin using 2,2'-azino-bis-(3-ethylbenzthiazoline-6-sulfonic acid) (ABTS) in the presence of peroxidase, *Anal. Biochem.* 185:335-338.

Aussenac, T., Lacombe, S., and Dayde, J., 1998, Quantification of isoflavones by capillary zone electrophoresis in soybean seeds: effects of variety and environment, *Am. J. Clin. Nutr.* 68S:1480S-1485S.

Barnes, S., Coward, L., Kirk, M., and Sfakianos, J., 1998, HPLC-mass spectrometry analysis of isoflavones, *Proc. Soc. Exp. Biol. Med.* 217:254-262.

Barnes, S., Kirk, M., and Coward, L., 1994, Isoflavones and their conjugates in soy foods: extraction conditions and analysis by HPLC-mass spectrometry, *J. Agric. Food Chem.* 42:2466-2474.

Barnes, S., Wang, C.-C., Smith-Johnson, A., and Kirk, M., 1999, Liquid chromatography: mass spectrometry of isoflavones, *J. Medicinal Food* 2:111-117.

Berhow, M. A., and Vaughn, S. F., 1999, Higher plant flavonoids: biosynthesis and chemical ecology, In: *Principles and Practices in Plant Ecology: Allelochemical Interactions.* Inderjit, Dakshini, K. M. M., and Foy, C., eds., CRC Press, Boca Raton, FL, pp. 423-438.

Bocchini, P., Russo, M., and Galletti, G. C., 1998, Pyrolysis-gas chromatography/mass spectrometry used as a microanalytical technique for the characterization of *Origanum heracleoticum* from Calabria, southern Italy, *Rapid Commun. Mass Spectrom.* 12:1555-1563.

Cancalon, P. F., 1999, Analytical monitoring of citrus juices by using capillary electrophoresis, *J. AOAC Int.* 82:95-106.

Careri, M., Elviri, L., and Mangia, A., 1999, Validation of a liquid chromatography ion spray mass spectrometry method for the analysis of flavanones, flavones and flavonols, *Rapid Commun. Mass Spectrom.* 13:2399-2405.

Careri, M., Mangia, A., and Musci, M., 1998, Overview of the applications of liquid chromatography-mass spectrometry interfacing systems in food analysis: naturally occurring substances in food, *J. Chromatog. A* **794**:263-297.

Colegate, S. M., and Molyneux, R. J., 1993, *Bioactive Natural Products: Detection, Isolation, and Structural Determination*, CRC Press, Boca Raton, FL.

Cordell, G. A., 1995, Changing strategies in natural products chemistry, *Phytochem.* **40**:1585-1612.

Coward, L., Smith, M., Kirk, M., and Barnes, S., 1998, Chemical modification of isoflavones in soyfoods during cooking and processing, *Am. J. Clin. Nutr.* **68**S:1486S-1491S.

Creaser, C. S., Koupai-Abyazani, M. R., and Stephenson, G. R., 1992, Gas chromatographic-mass spectromic characterization of flavanones in citrus and grape juices, *Analyst* **117**:1105-1109.

Dakora, F. D., 1995, Plant flavonoids: biological molecules for useful exploitation, *Austral. J. Plant Physiol.* **22**:87-99.

Davis, W. B., 1947, Determination of flavanones in citrus fruits, *Anal. Chem.* **19**:467-478.

Deng, H., and Van Berkel, G. J., 1998, Electrospray mass spectrometry and UV/Visible spectrophotometry studies of aluminum(III)-flavonoid complexes, *J. Mass Spectrom.* **33**:1080-1087.

Dixon, R. A., 1999, Isoflavonoids: biochemistry, molecular biology, and biological functions, In: *Comprehensive Natural Products Chemistry*, Sankawa, U., ed., Elsevier, New York, NY, pp. 773-823.

Dixon, R. A., and Paiva, N. L., 1995, Stress-induced phenylpropanoid metabolism, *Plant Cell* **7**:1085-1097.

Garcia, M. C., Torre, M., Marina, M. L., and Laborda, F., 1997, Composition and characterization of soyabean and related products, *Crit. Rev. Food Sci. Nutr.* **34**:361-391.

Gengross, O., and Renda, N., 1966, Occurrence and quantitative estimation of naringin in citrus juices, *Justus Liebigs Ann. Chem.* **691**:186-189.

Hahlbrock, K., and Scheel, D., 1989, Physiology and biochemistry of phenylpropanoid metabolism, *Ann Rev Plant Physiol. Plant Mol Biol.* **40**:347-369.

Harborne, J. B., Mabry, T. J., and Mabry, H., eds., 1975, *The Flavonoids*, Chapman & Hall, New York, NY and London, UK.

Harborne, J. B., and Mabry, T. J., eds., 1982, *The Flavonoids: Advances in Research*, Chapman & Hall, New York, NY and London, UK.

Harborne, J. B., ed., 1988, *The Flavonoids: Advances in Research since 1980*, Chapman & Hall, New York, NY and London, UK.

Harborne, J. B., ed., 1994, *The Flavonoids: Advances in Research since 1986*, Chapman & Hall, New York, NY and London, UK.

Hasegawa, S., Berhow, M. A., and Fong, C. H., 1995, Analysis of bitter principles in *Citrus*, In: *Modern Methods of Plant Analysis, Volume 18: Fruit Analysis*, Linskens, H.F. and Jackson J. F., eds., Springer Verlag, Berlin; Heidelberg, pp. 59-80.

Heinonen, S., Wahala, K., and Adlercreutz, H., 1999, Identification of isoflavone metabolites dihydrodaidzein, dihydrogenistein, 6'-OH-O-DMA, and *cis*-4-OH-equol in human urine by gas chromatography-mass spectroscopy using authentic reference compounds, *Anal. Biochem.* **274**:211-9.

Heller, W., and Forkmann, G., 1994, Biosynthesis of flavonoids, in: *The Flavonoids*, Harborne, J. B., ed., Chapman & Hall, New York, NY, USA, pp. 499-536.

Hendrickson, R., Kesterson, J. W., and Edwards, G. J., 1958, Ultraviolet absorption technique to determine the naringin content of grapefruit, *Proc. Fla. State Hort. Soc.* **71**:194-198.

Horowitz, R. M., 1957, Detection of flavanones by reduction with sodium borohydride, *J. Org. Chem.* **22**:1733-1734.

Jourdan, P. S., Mansell, R. L., Oliver, D. G., and Weiler, E. W., 1984, Competitive solid phase enzyme-linked immunoassay for the quantification of limonin in citrus, *Anal. Biochem.* **138**:19-24.

Jourdan, P. S., Mansell, R. L., and Weiler, E. W., 1982, Radioimmunoassay for the citrus bitter principle, naringin, and related flavonoid-7-*O*-neohesperidosides, *J. Medicinal Plant Res.* **44**:82-86.

Jourdan, P. S., McIntosh, C. A., and Mansell, R. L., 1985a, Naringin levels in citrus tissues. II. Quantitative distribution of naringin in *Citrus paradisi* Macfad., *Plant Physiol.* **77**:903-908.

Jourdan, P. S., Weiler, E. W., and Mansell, R. L., 1985b, Naringin levels in citrus tissues. I. Comparison of different antibodies and tracers for the radioimmunoassay of naringin, *Plant Physiol.* **77**:896-902.

Jungbuth, G., and Ternes, W., 2000, HPLC separation of flavanols, flavones and oxidized flavanols with UV-, DAD-, electrochemical and ESI-ion trap MS detection, *Fresenius J. Anal. Chem.* **367**:661-666.

Kirchner, J. G., ed., 1978, *Thin-Layer Chromatography*, Second Edition, Techniques of Chemistry Series Vol. XIV, John Wiley & Sons, New York, NY.

Kohen, F., Lichter, S., Gayer, B., DeBoever, J., and Lu, L.. J., 1998, The measurement of the isoflavone daidzein by time resolved fluorescent immunoassay: a method for assessment of dietary soya exposure, *J. Steroid Biochem. Mol Biol.* **64**:217-222.

Kudou, S., Fleury, Y., Wetli, D., Magnolato, D., Uchida, T., Kitamura, K. and Okubo, K., 1991, Malonyl isoflavone glycosides in soybean (*Glycine max* Merrill), *Agric. Biol. Chem.* **55**:2227-2233.

Kwierty, A., and Braverman, J. B., 1959, Critical evaluation of the cyanidin reaction for flavonoid compounds, *Bull. Res. Council Israel Sect. C* 7:187-196.

Lin, L.-Z., He, X.-G., Lindenmaier, M., Yang, J., Cleary, M., Qiu, S.-X., and Cordell, G. A., 2000, LC-ESI-MS study of the flavonoid glycoside malonates of red clover (*Trifolium pratense*). *J. Agric. Food Chem.* 48:354-365.

Luthria, D. L., Jones, A. D., Donovan, J. L., and Waterhouse, A. L., 1998, GC-MS determination of catechin and epicatechin levels in human plasma, Book of Abstracts, 215th ACS National Meeting, Dallas, March 29-April 2, 1998. AGFD 008.

Mabry, T. J., Markham, K. R., and Thomas, M. B., 1970, *The Systematic Identification of Flavonoids,* Springer Verlag, New York, NY.

Mann, J., R. Davidson, S., Hobbs, J. B., Banthorpe, D. V., and Harborne, J. B., 1994, *Natural Products: Their Chemistry and Significance,* Addison Wesley Longmann Ltd., Edinburgh Gate, Harlow.

Matsumoto, R. and Okudai, N., 1991, Early evaluation of citrus bitter component, flavanone neohesperidosides by enzyme immunoassay using anti-naringin antibody, *J. Japanese Soc. Horticult. Sci.* 60:191-200.

Messina M. J., 1999, Legumes and soybeans: overview of their nutritional profiles and health effects, *Am. J. Clin. Nutr.* 70S:439S-450S.

Novotny, L., Vachalkova, A., Al-Nakib, T., Mohanna, N., Vesela, D., and Suchy, V., 1999, Separation of structurally related flavonoids by GC/MS technique and determination of their polarographic parameters and potential carcinogenicity, *Neoplasma* 46:231-236.

Pietta, P. G., Mauri, P. L., Rava, A., and Sabbatini, G., 1991, Application of micellar electrokinetic capillary chromatography to the determination of flavonoid drugs, *J. Chromatog. A* 549:367-373.

Runkel, M., Muehlau, A., Duecker, D., Tegtmeier, M., and Legrum, W., 1998, Capillary electrophoresis (CE): An efficient tool for the quality control of fruits as shown for the constituents of grapefruit, *Fruit Process.* 8:102-104.

Seigler, D. S., 1981, Secondary metabolites and plant systematics, In: *Secondary Plant Products,* Stumpf, P. K., and Conn, E. E., eds., Academic Press, New York, pp. 139-175.

Seitz, U., Oefner, P. J., Nathakarnkitkool, S., Popp, A., and Bonn, G. K., 1992, Capillary electrophoretic analysis of flavonoids, *Electrophoresis* 13:35-38.

Shelnutt, S. R., Cimino, C. O., Wiggins, P. A., and Badger, T. M., 2000, Urinary pharmacokinetics of the glucuronide and sulfate conjugates of genistein and daidzein, *Cancer Epidemiol. Biomarkers Prev.* 9:413-419.

Shihabi, Z. K., Kute, T., Garcia, L. L., and Hinsdale, M., 1994, Analysis of isoflavones by capillary electrophoresis, *J. Chromatog. A* 680:181-185.

Song, T., Barua, K., Buseman, G., and Murphy, P. A., 1998, Soy isoflavone analysis: quality control and a new internal standard, *Am. J. Clin. Nutr.* 68S:1474S-1479S.

Stobiecki, M., 2000, Application of mass spectrometry for identification and structural studies of flavonoid glycosides. *Phytochem.* 54:237-256.

Stobiecki, M., Malosse, C., Kerhoas, L., Wojlaszek, P., and Einhorn, J., 1999, Detection of isoflavonoids and their glycosides by liquid chromatography/electrospray ionization mass spectrometry in root extracts of lupin (*Lupinus albus*), *Phytochem. Anal.* 10:198-207.

Swantsitang, P., Tucker, G., Robards, K., and Jardine, D., 2000, Isolation and identification of phenolic compounds in *Citrus sinensis, Anal. Chim Acta* 417:231-240.

Tsukamoto, C., Shimada, S., Igata, K., Kudou, S., Kokubun, M., Okubo, K., and Kitamura, K., 1995, Factors affecting isoflavone content in soybean seeds: changes in isoflavones, saponins, and composition of fatty acids at different temperatures during seed development, *J. Agric. Food Chem.* 43:1184-1192.

Voirin, B., Sportouch, M., Raymond, O., Jay, M., Bayet, C., Dangles, O., and El Hajji, H., 2000, Separation of flavone *C*-glycosides and qualitative analysis of *Passiflora incarnata* L. by capillary zone electrophoresis, *Phytochem. Anal.* 11:90-98.

Wang, C. Y., Ma, Q., Pagadala, S., Sherrard, M. S., and Krishnan, P. G., 1998, Changes of isoflavones during processing of soy protein isolates, *J. Amer Oil Chem. Soc.* 75:337-342.

Wang, C. Y., Sherrard, M., Pagadala, S., Wixon, R., and Scott, R. A., 2000, Isoflavone content among maturity group 0 to 11 soybeans, *J. Amer Oil Chem. Soc.* 77:483-487.

Wang, H., and Murphy, P. A., 1994, Isoflavone content in commercial soybean foods, *J. Agric. Food Chem.* 42:1666-1673.

Wang, H. J., and Murphy, P. A., 1996, Mass balance study of isoflavones during soybean processing, *J. Agric. Food Chem.* 44:2377-2383.

Watson, D. G., and Pitt, A. R., 1998, Analysis of flavonoids in tablets and urine by gas chromatography/mass spectrometry and liquid chromatography/mass spectrometry, *Rapid Commun. Mass Spectrom.* 12:153-156.

Wolfender, J. L., Rodriguez, S., Hostettmann, K., and Wagner-Redecker, W., 1995, Comparison of liquid chromatography/electrospray, atmospheric pressure chemical ionization, thermospray and continous flow fast atom bombardment mass spectrometry for the determination of secondary metabolites in crude plant extracts, *J. Mass Spectrom. and Rapid Commun. Mass Spectrom. Special:* S35-S46.

Zhou, S,. and Hamberger, M., 1996, Application of liquid chromatography atmospheric pressure ionization mass
    spectrometry in natural products analysis: Evaluation and optimization of electrospray and heated nebulizer
    interfaces, *J. Chromatog.* A **755**:189-204.

HPLC-MASS SPECTROMETRY OF ISOFLAVONOIDS IN SOY AND THE AMERICAN GROUNDNUT, *APIOS AMERICANA*

S. Barnes[1,2], C-C. Wang[2], M. Kirk[2], M. Smith-Johnson[1], L. Coward[2], N. C. Barnes[3], G. Vance[4], and B. Boersma[1]

1. INTRODUCTION

There is an ever growing interest in the study of the polyphenols. Their importance in the prevention of chronic disease is gradually being unfolded. Examples range from those such as the flavonoid quercitin and the stilbene resveratrol in red wine (Das et al., 1999) to the isoflavonoids daidzein, genistein and glycitein in soy (Barnes, 1998).

Despite the intensity of the UV absorbance of the polyphenols, in particular the bioflavonoids, their measurement in biological fluids and tissues requires more sensitive and specific techniques than can be provided by HPLC combined with diode array UV-Vis absorption (Barnes et al., 1998b; Barnes et al., 1999; Merken et al., 2000). Furthermore, metabolism of bioflavonoids leads to products that have much lower molar UV absorbance than their parent compounds. Such metabolites are therefore difficult to detect by this method.

Mass spectrometry provides both the sensitivity and the specificity needed for the analysis of bioflavonoids. There have been many applications of gas chromatography-mass spectrometry methods to the analysis of isoflavonoids and other polyphenols. However, as discussed in previous reviews of this topic (Barnes et al., 1998b; Barnes et al., 1999), the combination of HPLC with mass spectrometry (LC-MS) greatly simplifies these measurements, because it obviates the need for preparation of volatile derivatives and makes much of the workup procedures unnecessary. In a recent review the rationales for the selection of the type of mass spectrometer instrument were also presented (Barnes et al., 1999).

---

[1] Department of Pharmacology & Toxicology and [2]Comprehensive Cancer Center Mass Spectrometry Shared Facility, University of Alabama at Birmingham, Birmingham, AL 35294. [3]The Resource Learning Center, Shades Valley High School, Birmingham, AL 35209. [4]Vestavia High School, Vestavia, AL 35216

Flavonoids in Cell Function, edited by Béla Buslig and John Manthey.
Kluwer Academic/Plenum Publishers, New York, 2002.

In this present study we discuss (1) methods for the identification of specific isoflavonoids, particularly in their conjugated forms, using LC-MS, (2) optimization of methods for quantitative measurements of isoflavonoids, and (3) approaches to substantially increase the sensitivity of the measurement of isoflavonoids. Each of these issues are relevant to the investigation of other polyphenols. In a new application of these methods, LC-MS has been used to study the isoflavonoids in tubers of the American groundnut, *Apios americana*.

2. MATERIALS & METHODS

2.1. Materials

Isoflavone and their metabolites standards were either isolated from soy molasses (Peterson and Barnes, 1991), purchased from Indofine (Somerville, NJ), or were gifts from Kristiina Wähälä, University of Helsinki, Finland. Tubers of *Apios americana* were provided by Dr. Bill Blackmon, Dr. Jackie Carlusi-Dunlop and colleagues at the University of Southwestern Louisiana. Soy food samples were purchased at a local store specializing in vegetarian foods. Human plasma and urine samples were obtained in a study of the effect of soy protein on plasma biomarkers in a group of elderly men (Smith et al., 1999). This study was performed under a protocol approved by the UAB Institutional Human Use Review Committee. Plasma and urine samples from female rats were provided by Dr. Clinton Grubbs (University of Alabama at Birmingham) from a study funded by a contract from the National Cancer Institute and approved by the Institutional Animal Use and Care Committee.

2.2. Methods

Soy food samples were extracted in ten volumes of ice-cold 80% aqueous methanol by tumbling for 2 h at 4°C. This method is essential to prevent the hydrolysis of 6"-*O*-malonyl and acetyl esters of isoflavonoid β-glycosides (Coward et al., 1998).

Plant specimens were homogenized at room temperature in four volumes of methanol. The homogenates were filtered and the residues re-extracted with 80% aqueous methanol. Extracts were combined and taken to dryness using a rotary evaporation at 30°C. The dried residues were dispersed in a minimum volume of 80% aqueous methanol and extracted three times with five volumes of n-hexane to remove triglycerides. The methanolic phase was evaporated to dryness and redissolved in 80% aqueous methanol prior to HPLC and LC-MS analysis.

To measure total (free and conjugated) isoflavonoids and their metabolites in plasma and urine, specimens were hydrolyzed with β-glucuronidase/sulfatase and the aglucones recovered by ether extraction, as described previously (Coward et al., 1996). Conjugated isoflavonoids were isolated by their absorption to and subsequent elution from a C_8 Sep-Pak cartridge (Coward et al., 1996).

Mass spectrometry analyses were performed with a PE-Sciex (Concord, Ontario, Canada) API III triple quadrupole instrument using commercial electrospray ionization (IonSpray™) or heated nebulizer atmospheric pressure chemical ionization (HN-APCI) interfaces, or a hand-built nanoelectrospray ionization interface. Mass spectra were recorded in both the positive and negative modes. Daughter ion MS-MS spectra were recorded following selection of specific parent molecular ions and collision-induced dissociation with Argon-N_2 gas (90:10, by vol).

HPLC analyses were performed on a Hewlett Packard model 1050 (for LC-MS) or a model 1100 for diode array UV-vis analyses. The reverse-phase C_8 column (25 cm x 4.6 mm i.d. for analyses with the HN-APCI interface and 10 cm x 2.1 mm i.d. for analysis with the Ion Spray™ interface) was eluted with a linear gradient of 0-40% solvent B (95% acetonitrile in 10 mM ammonium acetate, pH 6.5) in solvent A (10% acetonitrile in 10 mM ammonium acetate, pH 6.5). The flow rate in the analyses using the HN-APCI interface was 1.5 mL/min, whereas it was 0.2 mL/min using the Ion Spray™ interface. In the latter case, the eluate was split into two and 100 μL/min passed to the Ion Spray™ interface. The nanoelectrospray interface was constructed by treating bared quartz capillary tubing with HF to produce a tapered end. Electrical contact was achieved by joining a stainless steel capillary to the quartz tip using a teflon sleeve. Samples were introduced into the interface by flow injection at 200 nl/min produced by a Harvard infusion pump. The quartz tip was positioned off axis within 1 cm of the orifice of the API III. The voltage applied to the interface was -1400 V. Negative ion spectra were recorded using multiple reaction ion monitoring using a combination of parent molecular ions and specific daughter ions (see Coward et al., 1996).

3. RESULTS

3.1. Identification of Isoflavones

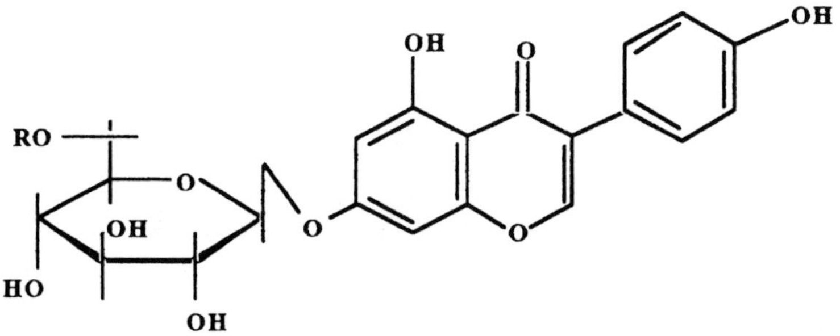

Figure 1. Genistein β-glucosides in soy foods. R=H, CH_3CO, and $HOOC.CH_2.CO$ for genistin, 6"-*O*-acetylgenistin and 6"-*O*-malonylgenistin, respectively.

Soy contains three isoflavones that are conjugated to form β-glucosides (Fig. 1). However, the major conjugated form found in the soybean is the 6"-malonyl ester of each β-glucoside (6MalGlc) (Fig. 1). This conjugate is broken down during processing of soy to form soy food products (Barnes, et al., 1998a; Barnes et al.,1994) and during cooking itself (Coward et al., 1998). Treatment of soybeans with pressurized boiling water is used to make soy milk and eventually tofu. This converts most of the 6MalGlc to the non-esterified β-glucosides (Barnes et al., 1994). Extraction of the soybean with hexane to recover soybean oil leaves behind a defatted soy flour product. This largely retains the composition of the isoflavone conjugates in soybeans (6MalGlc predominating). However, for animal feed it is first toasted to destroy residual lipoxygenase and protease activities. Heating leads to the decarboxylation of the malonyl group to form an acetyl group (Fig. 1). Thus the 6"-*O*-acetyl-β-glucosides (6AcGlc) are artificial, not being present in the soybean. However, this means that consumer soy products are often a complex mixture of isoflavones - for each isoflavone it may be in three conjugate forms as well as the aglucone (unconjugated form).

Isoflavones can be analyzed by reverse-phase HPLC using a linear gradient of acetonitrile in 0.1% v/v trifluoroacetic acid (Coward et al., 1993) or 1-5% v/v acetic acid (Wang et al., 1994). However, there is overlap between the various conjugate classes in both these solvent systems. In Fig. 2, the chromatograms of soy flour (A) versus soy germ (B) show marked differences in composition. Daidzein and glycitein are found in much larger amounts than genistein in soy germ as compared to whole soy products.

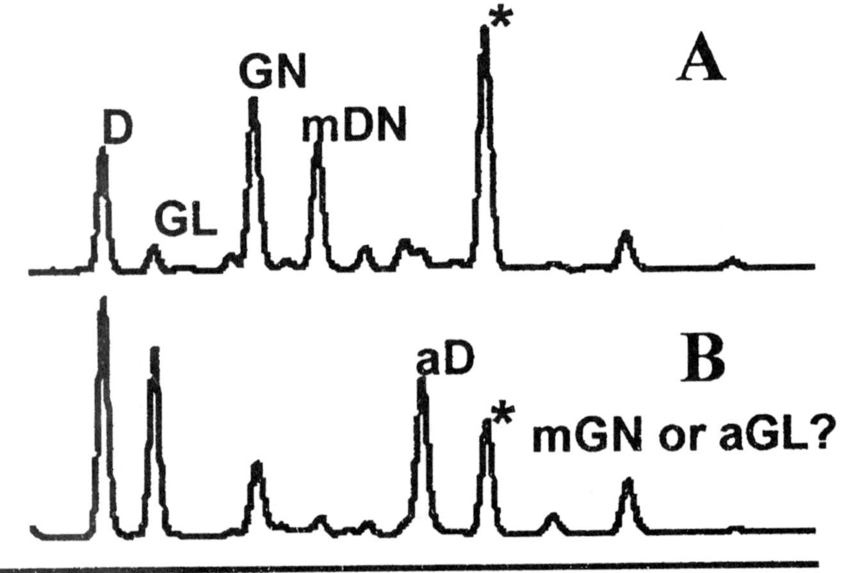

Figure 2. Reverse-HPLC analysis of isoflavone glucoside conjugates in soy protein isolate (A) and a soy germ food product (B). Identity of the isoflavones: D, daidzin; GL, glycitin; GN, genistin; mDN, 6"-*O*-malonyl-daidzin; mGN, 6"-*O*-malonylgenistin; aD, 6"-*O*-acetyldaidzin. The peak eluting at the same time as mGN was another isoflavone (see Figures 3-5). The isoflavones were separated using a 0-40% linear gradient of acetonitrile in 0.1% trifluoroacetic acid on a 22 cm x 4.6 mm i.d. C_8 reverse-phase column.

The peak in the soy germ extract marked with an asterisk in Fig. 2 had a retention time of either daidzein 6AcGlc or genistein 6MalGlc. From the UV spectra of the corresponding peaks from soy isoflavones and soy germ, the soy germ peak was identified as daidzein 6AcGlc (Fig. 3).

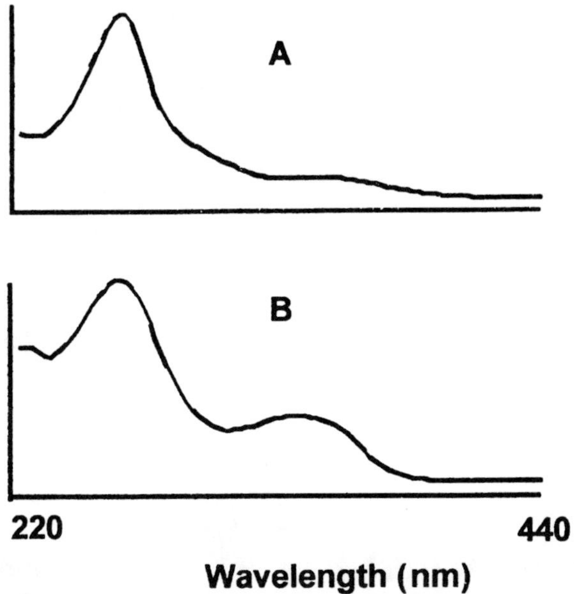

Figure 3. The unknown HPLC peak in the soy germ food is not 6"-*O*-malonylgenistin. UV spectra of the corresponding peaks in soy protein isolate (A) and soy germ food (B) are consistent with genistein and glycitein, respectively.

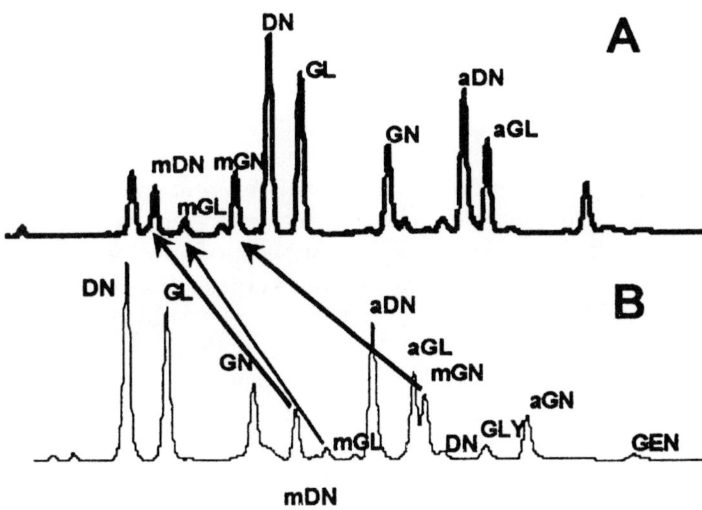

Figure 4. The effect of changing the pH of the mobile phase during reverse-phase HPLC of isoflavones. A mixture of extracts of soy protein isolate and a soy germ food were analyzed using an acetonitrile gradient in 10 mM ammonium acetate, pH 7.0 (A) and 0.1% trifluoroacetic acid (B). The malonyl esters of the isoflavones shifted from the middle of the chromatogram (in B) to the front (in A).

In trifluoroacetic acid or acetic acid, the malonate group of 6MalGlc conjugates will be protonated. This increases its hydrophobic nature and slows the elution of 6MalGlc conjugates from the reverse-phase column. We therefore examined the effect of a neutral pH solvent (10 mM ammonium acetate) on chromatographic separation. As expected, the 6MalGlc derivatives became more hydrophilic and eluted in front of the β-glucosides (Fig. 4A). Indeed, complete class separation was achieved in this solvent system.

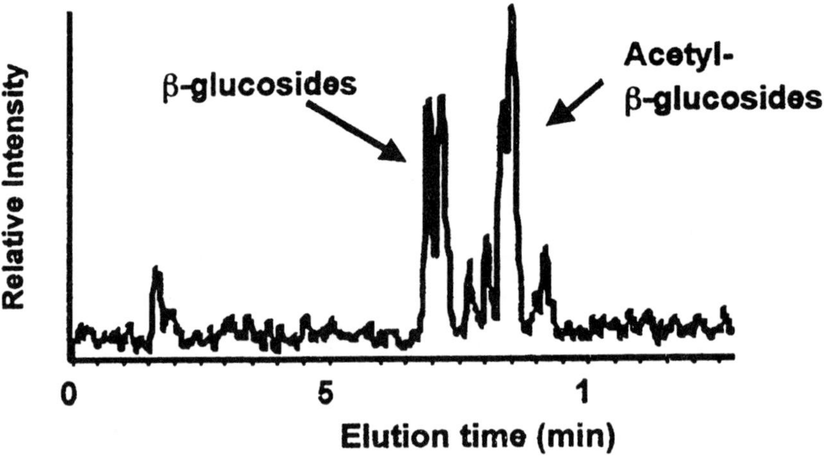

Figure 5. Isoflavone conjugates in extracts of a soy germ food product separated by HPLC-ESI-MS. This ion chromatogram is for all ions with *m/z* values above 400.

The neutral solvent system is highly suited to reverse-phase HPLC-electrospray ionization-mass spectrometry. The isoflavones and their conjugates readily form negatively or positively charged molecular ions. Under the conditions needed to remove the adducts with the solvent components, the molecular ion will fragment with loss of the glucoside group to yield the aglucone ion, a useful confirmation of identity as an isoflavone. Using the soy germ material as an example, an ion chromatogram (for ions with *m/z* values above 400) can be used to display the conjugated isoflavones in a simple aqueous methanol extract (Fig. 5). The peaks do not seem well separated because the selectivity of the mass spectrometer detector allows the use of very fast solvent gradients. By using specific molecular ions for each isoflavone conjugate, selected ion chromatograms can be then be created from the primary data set (Fig. 6). Each isoflavone conjugate can be readily seen as a single, sharp peak. Confirmation of identification of the isoflavone conjugates can be carried out by selecting the molecular ion or the aglucone ion and producing daughter fragment ions by collision-induced dissociation with Argon-N_2 gas (see below).

Following our initial work on the isoflavonoids in soy, we turned our attention to the tubers of *A. americana*. Simple methanol extraction of these potato-like vegetables and analysis by reverse-phase HPLC revealed that there was a large amount of a UV-absorbing substance and many other minor UV absorbing peaks. Treatment with β-glucosidase caused most of the peaks to shift to longer retention times, suggesting that they were β-glucosides. Using chromatography on Sephadex LH-20, the major peak (in its intact form) was isolated. ESI-MS analysis (m/z 595 for the [M+H]$^+$ molecular ion, m/z 433 for [M+H-162]$^+$ and m/z 271 for the aglucone ion) and proton NMR revealed that this compound was a diglucoside of genistein (Barnes, 1993). The site of attachment of the glucose residues was the 7-position, identifying this conjugate as a 7-O-glucosylglucoside of genistein.

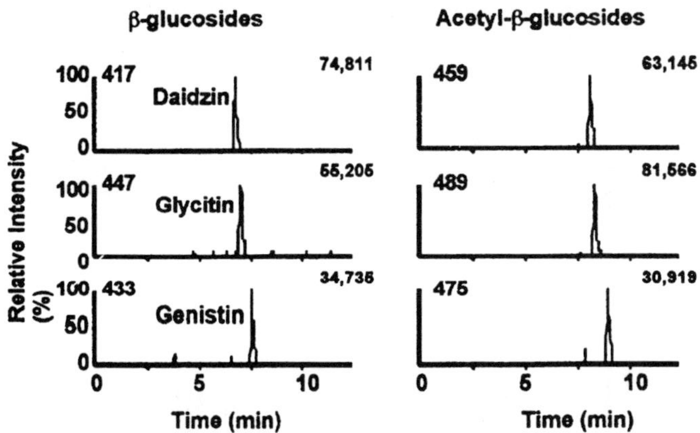

Figure 6. Selected ion chromatograms following HPLC-ESI-MS of extracts of a soy germ food product. The values in the top left of each chromatogram are the m/z values of the [M+H]$^+$ molecular ions. The values in the top right of each chromatogram are the maximum ion intensities.

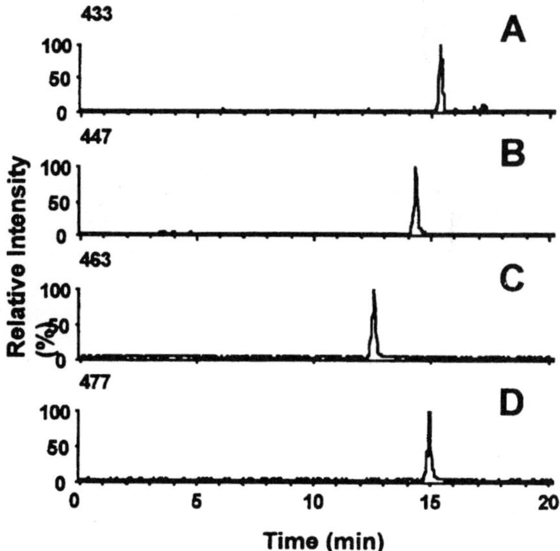

Figure 7. LC-ESI-MS of isoflavone glycosides in tubers of *A. americana*. Defatted methanol extracts were separated on a 0-30% gradient of acetonitrile in 0.1% aqueous acetic acid. Positive ion spectra were collected. Data are presented as selected ion chromatograms for monoglucosides of genistein (A), a tetrahydroxyisoflavone (B), a methoxy-tetrahydroxylated isoflavone (C), and a methoxy-pentahydroxylated isoflavone (D).

Unlike genistin, the β-glucoside of genistein, the glucosylglucoside was freely water soluble. Other peaks were isoflavones with *m/z* values consistent with the addition of 1-2 additional hydroxyl groups as well as methoxy groups, as shown in the selected ion chromatogram (Fig. 7).

3.2. Quantitative Measurements

LC-MS analysis using a triple quadrupole mass spectrometer provides both high sensitivity and high specificity. As in MS-MS experiments, the analytical method depends

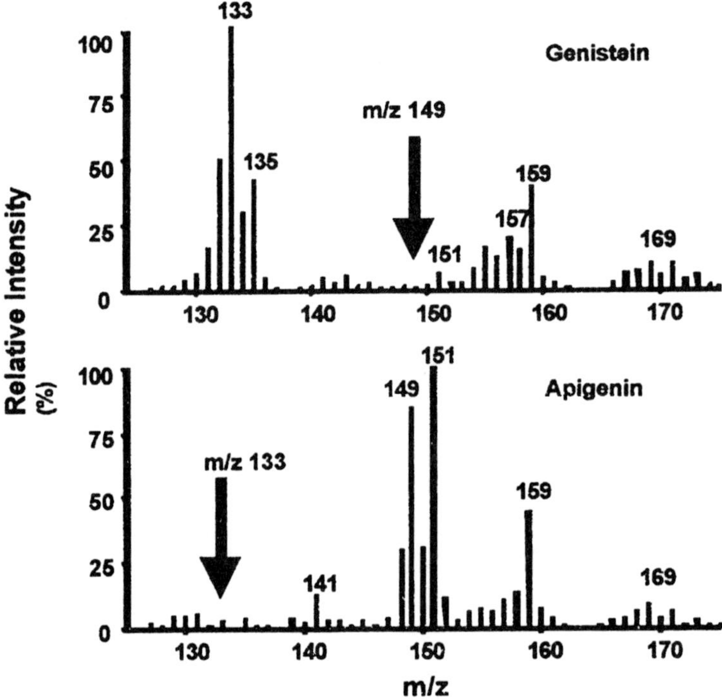

Figure 8. Expanded view of the negative daughter ion mass spectra of genistein and apigenin. The *m/z* 269 [M-H]- molecular ions of genistein and its flavonoid isomer apigenin were fragmented by collision with Ar:N₂ (90:10). Unique daughter ions (*m/z* 133 for genistein; *m/z* 149 for apigenin) were identified. [published with the permission of the Journal of Medicinal Food].

on the selection of the parent molecular ion and its fragmentation by collision-induced dissociation to produce daughter ions. Careful selection of one or more daughter ions allows the specific measurement of an individual flavonoid in the presence of many other flavonoids and indeed other unrelated substances (Coward et al., 1996). Examination of the MS-MS spectra of genistein and its flavonoid isomer apigenin revealed that two daughter ions (*m/z* 133 for genistein and *m/z* 149 for apigenin) completely distinguish between these two compounds (Fig. 8) that otherwise have the same molecular weight (Barnes et al., 1999).

This technique is known as reaction ion monitoring. The mass spectrometer can rapidly switch between different pairs of parent ion/daughter ion combinations (multiple reaction ion monitoring - MRM) during an HPLC analysis. This technique greatly simplifies the analysis of isoflavones, particularly since it removes the necessity for gradient chromatography. Isoflavones can be analyzed over a period of just a few minutes without the need for re-equilibration of the HPLC column (Coward et al., 1996). In the examples shown in Fig. 9, the much higher levels of isoflavones in soy sauce prepared by fermentation as opposed to chemical hydrolysis can be readily seen.

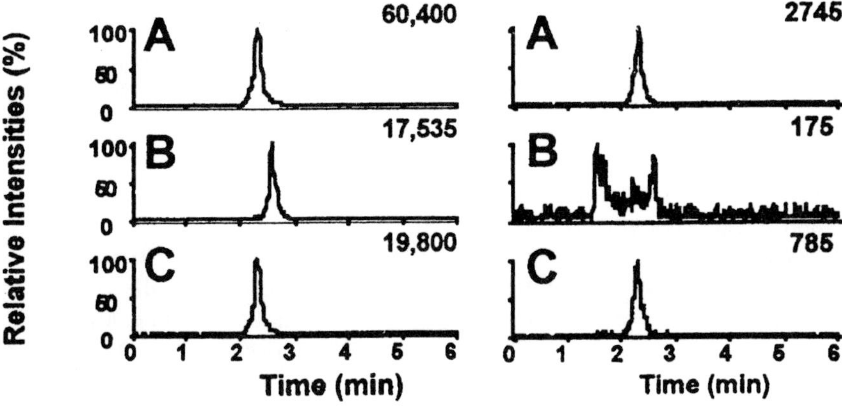

Figure 9. LC-multiple reaction ion monitoring-MS of isoflavonoids in commercial soy sauces. Unconjugated isoflavones in two soy sauces were recovered by solid-phase extraction and analyzed by LC-MRM-MS under isocratic conditions (30% acetonitrile in 10 mM ammonium acetate) on a 10 cm x 2.1 mm i.d. C_8 reverse-phase column. The isoflavones analyzed were daidzein (253/223), genistein (269/133) and glycitein (283/240). The numbers on the top right corner of each chromatogram represent the full scale value of the ion intensity. Ion chromatograms on the left are from a fermented soy sauce, whereas those on the right are from a chemically hydrolyzed soy sauce.

Because of the high selectively of the MRM procedure, and the use of isocratic conditions, large number of samples can be analyzed in a session (Coward et al., 1996). The reproducibility of the MRM procedure for isoflavones is in the range of 3-8% (expressed as the coefficient of variation - the standard deviation of duplicate samples divided by their mean) for concentrations in a 1 mL sample of 40 nM or greater. The limit of detection is in the range from 2-10 nM, depending on the intensity of the molecular ion and the resulting selected daughter ion (Barnes et al., 1999).

3.3. High Sensitivity

Future improvements in the sensitivity of the measurement of isoflavones and all other polyphenols are needed. In our present procedure, using conventionally sized columns (2.1 or 4.6 mm i.d.), the column effluent (0.2-1 mL/min) is split with 100 µL/min being passed to the ESI interface. This flow rate rather inefficiently transfers solutes from the liquid to the gas phase. Greater transfer efficiency can be achieved by lowering the flow rate through the ESI interface. We have built an interface which utilizes a flow rate of 200 nl/min. At this flow rate, isoflavone concentrations as low as 2-3 nM can be detected. This implies the amounts of sample injected on a suitable column can be as low as 400-600 amol. We have developed a capillary membrane procedure that is able to measure genistein concentrations as low as 1-2 nM in a 1 μL sample (Wang et al., 2000). However, developing an HPLC procedure for this purpose remains a considerable challenge.

4. DISCUSSION

Even though many isoflavones and other polyphenols have strong UV absorbance, LC-MS is rapidly becoming the method of choice for the analysis of these compounds and their metabolites (Barnes et al., 1998a; Cimino et al., 1999; Holder et al., 1999; Stevens et al., 1999; Doerge et al., 2000; Shelnutt et al., 2000). Mass spectrometers are more universal than other methods of analyte detection. Moreover, the mass selection of the parent molecular ions and of daughter ions formed by gas-phase collisions provides not only great sensitivity but also specificity.

Having the ability to fragment ions is a powerful tool in mass spectrometry (Barnes et al., 1998b; Coldham et al., 1999). It is important when studying isoflavones and other polyphenols to confirm the identity of each peak detected. For example, it is particularly useful for the measurement of newly discovered isoflavone metabolites such the brominated (Boersma et al., 2000) and chlorinated and nitrated isoflavones (Boersma et al., 1999). By using selected parent and daughter ion combinations, LC-MS enables accurate and specific measurements of individual compounds in complex matrices. At the present time the sensitivity of this method has not been fully exploited. Using conventional HPLC columns and electrospray ionization, only a fraction of the eluate is taken for analysis. Even then the spraying conditions (μL/min flow rates) are inefficient with less than 1% of the ions formed actually entering the mass spectrometer (Geromanos et al., 2000). Slowing the flow rate into the nl/min range increases the ionization/delivery efficiency. For peptides, we have achieved increases in sensitivity of 10-20 fold for the same solution by spraying at 200 nl/min as opposed to 10 μL/min (Wang, C-C., Kirk, M. and Barnes, S., University of Alabama at Birmingham, unpublished observations). The challenge is how to carry out chromatography at 100-300 nl/min and not have to proportionately decrease the sample size (and thereby lose the advantage that would be gained). This problem is being dealt with by those analyzing peptides resulting from proteolytic digestion of proteins. The sample size issue is overcome by using gradient elution methods where the sample is loaded under conditions (low solvent) where tight binding is promoted. The bound compounds are eluted with a solvent gradient. However, there is a price to pay - at 200 nl/min, a 5 μL sample will take 25 min to load.

Unlike GC-MS, that has been widely used in the past (Morton et al., 1999), LC-MS can just as easily analyze conjugated isoflavones as it can the corresponding aglucones. Also, it does not require the formation of chemical derivatives. This ability proved to be very helpful in the identification of the glycosyl derivatives of the isoflavones in tubers of *A. americana*. This new source of isoflavone conjugates contains genistein in concentrations that for some strains exceeds those found in soy beans (Barnes ,S., Barnes, N. C., Blackmon, W., Carlusi-Dunlop, J., and Vance, G., University of Alabama at Birmingham and University of Southwestern Louisiana, unpublished observations). Interestingly, unlike soybeans, the diglucoside nature of the genistein conjugate in *A. americana* makes it noticeably more soluble and potentially more readily absorbable in the small intestine. The tubers of *A. americana* have been part of the diet of Eastern native Americans for many centuries and were eaten at the first Thanksgiving feasts by the pilgrims arriving from Europe and their hosts (Reynolds et al., 1990).

5. ACKNOWLEDGMENTS

These studies were supported in part by grants from the American Institute for Cancer Research (91B47R), the National Cancer Institute (5R01 CA-61668), the United Soybean Board (7312) and a Howard Hughes Medical Institute Infrastructure Support grant to the University of Alabama at Birmingham. The mass spectrometer was purchased by funds from a NIH Instrumentation Grant (S10RR06487) and from this institution. Operation of the UAB Comprehensive Cancer Center Mass Spectrometry Shared Facility has been supported in part by a NCI Core Research Support Grant to the UAB Comprehensive Cancer Center (P30 CA13148).

6. REFERENCES

Barnes, K. A., Smith, R.A., Williams, K., Damant, A.P., and Shepherd, M. J., 1998a, A microbore high performance liquid chromatography/electrospray ionization mass spectrometry method for the determination of the phytoestrogens genistein and daidzein in comminuted baby foods and soya flour, *Rapid Commun. Mass Spectrom.* **12**:130-138.

Barnes, N. C., 1993, The American groundnut, *Apios americana*, contains a glycoside of the antitumor agent genistein, North Central Alabama High School Science Fair, 1993.

Barnes, S., 1998, Evolution of the history of soy and genistein. *Proc. Soc. Exp. Biol. Med.* **217**:386-392.

Barnes, S., Coward, L., and Kirk, M., 1998b, HPLC-mass spectrometry analysis of isoflavones, *Proc. Soc. Exp. Biol. Med.* **217**:254-262.

Barnes, S., Kirk, M., and Coward, L. J., 1994, Isoflavones and their conjugates in soy foods: extraction conditions and analysis by HPLC-mass spectrometry, *Agric. Food Chem.* **42**:2466-2474.

Barnes, S., Wang, C-C., Smith-Johnson, M., and Kirk, M., 1999, Liquid chromatography-mass spectrometry of isoflavones, *J. Med. Food* **2**:111-117.

Boersma, B. J., Patel, R. P., Kirk, M., Darley-Usmar, V. M., and Barnes, S., 1999, Chlorination and nitration of soy isoflavones, *Arch. Biochem. Biophys.* **368**:265-275.

Boersma, B. J., Patel, R. P., Kirk, M., Muccio, D. D., Darley-Usmar, V. M., and Barnes, S., 2001, Bromination, chlorination and nitration of isoflavonoids, In: *Free Radicals in Foods: Chemistry, Nutrition and Health,* Morello, M.., ed., ACS Press, in press.

Cimino, C. O., Shelnutt, S. R., Ronis, M. J. J., and Badger, T. M., 1999, An LC-MS method to determine concentrations of isoflavones and their sulfate and glucuronide conjugates in urine, *Clin. Chim. Acta* **287**:69-82.

Coldham, N. G., Howells, L. C., Santi, A., Montesissa, C., Langlais, C., King, L. J., Macpherson, D. D., and Sauer, M. J., 1999, Biotransformation of genistein in the rat: elucidation of metabolite structure by product ion mass fragmentology, *J. Ster. Biochem. Mol. Biol.* **70**:169-184.

Coward, L., Barnes, N. C., Setchell, K. D. R., and Barnes, S., 1993, The antitumor isoflavones, genistein and daidzein, in soybean foods of American and Asian diets, *J. Agric. Food Chem.* **41**:1961-1967.

Coward, L., Kirk, M., Albin, N., and Barnes, S., 1996, Analysis of plasma isoflavones by reversed-phase HPLC-multiple reaction ion monitoring-mass spectrometry, *Clin. Chim. Acta,* **247**:121-142.

Coward, L., Smith, M,, Kirk, M, and Barnes, S., 1998, Chemical modification of isoflavones in soy foods during cooking and processing, *Am. J. Clin. Nutr.* **68**:1486S-1491S.

Das, D. K., Sato, M., Ray, P. S., Maulik, G., Engelman, R. M., Bertelli, A. A., and Bertelli, A., 1999, Cardioprotection of red wine: role of polyphenolic antioxidants, *Drugs Exp. Clin. Res.* **25**:115-120.

Doerge, D. R., Chang, H. C., Churchwell, M. I., and Holder, C. L., 2000, Analysis of soy isoflavone conjugation *in vitro* and in human blood using liquid chromatography-mass spectrometry, *Drug Metab. Disp.* **28**:298-307.

Geromanos, S., Freckleton, G., and Tempst, P., 2000, Tuning of an electrospray ionization source for maximum peptide-ion transmission into a mass-spectrometer, *Anal. Chem.* **72**:777-790.

Holder, C. L., Churchwell, M. I., and Doerge, D. R., 1999, Quantification of soy isoflavones, genistein and daidzein, and conjugates in rat blood using LC/ES-MS, *J. Agric. Food Chem.* **47**:3764-3770.

Merken, H. M., and Beecher, G. R., 2000, Measurement of food flavonoids by high-performance liquid chromatography: A review, *J. Agric. Food Chem.* **48**:577-599.

Morton, M., Arisaka, O., and Miyake A., 1999, Analysis of phyto-oestrogens by gas chromatography-mass spectrometry, *Environ. Toxicol. Phar.* **7**:221-225.

Peterson, T. G., and Barnes, S., 1991, Genistein inhibition of the growth of human breast cancer cells: independence from estrogen receptors and the multi-drug resistance gene, *Biochem. Biophys. Res. Commun.* **179**:661-667.

Reynolds, B. D., Blackmon, W. J., Wickremesinhe, E., Wells, M. H., and Constantin, R. J., 1990, Domestication of *Apios americana*, In *Advances in New Crops*, Janick, J., and Simon, J. E., eds., Timber Press, Portland, OR, pp. 436-442.

Shelnutt, S. R., Cimino, C. O., Wiggins, P. A, and Badger, T. M., 2000, Urinary pharmacokinetics of the glucuronide and sulfate conjugates of genistein and daidzein, *Cancer Epidem. Biomark. Prevent.* **9**:413-419.

Smith, M., Kirk, M., Weiss, H., Irwin, W., Urban D., Grizzle, W. E., and Barnes, S., 1999, Serum and urinary isoflavonoids and their metabolites in elderly men on diets supplemented with beverages containing untreated and alcohol-extracted soy protein, *J. Med. Food* **2**:219-222.

Stevens, J. F., Taylor, A.W., and Deinzer, M.L., 1999, Quantitative analysis of xanthohumol and related prenyl-flavonoids in hops and beer by liquid chromatography-tandem mass spectrometry, *J. Chromatogr. A* **832**:97-107.

Wang, C. C., Kirk, M., Smith, M., and Barnes, S., 2000, A highly sensitive method for the analysis of isoflavones in microliter samples of physiological fluids by nanoelectrospray mass spectrometry (nanoES-MS), 48th Natl. Mtg. Am. Soc. Mass Spectrom., *Drugs and New Metabolism*, 132 (abs.).

Wang, H-J., and Murphy, P. A., 1994, Isoflavone content in commercial soybean foods, *J. Agric. Food Chem.* **42**:1666-1673.

HISTORY AS A TOOL IN IDENTIFYING "NEW" OLD DRUGS

John M. Riddle[1]

1. SAW PALMETTO, STINGING NETTLE AND AFRICAN PRUNE

When Merck released a new drug, finasteride, with the proprietary name of Proscar® for benign prostatic hypertrophy (BPH), they were surprised, delighted even, when they learned that users reported the promotion of hair growth. Soon they filed with the FDA for the release of the same drug (although lower dosage) with a new name, Propecia®, for treating alopecia or male pattern baldness. Often the West makes its discoveries seemingly serendipitously or, less frequently, through insightful hunches. An example of the latter is the brilliant research of Jean Wilson, and Donald Coffey and his collaborators, who discovered a synthetic intracellular enzyme that converts testosterone into an active androgen, a reductase inhibitor for treatment of BPH (Horton, 1992; Wilson, 1972). These modern endocrinologists and chemists apparently did not know that ancient peoples knew the same action (not necessarily the same mechanism of action) which they found in nettle (*Urtica dioica* L.) and saw palmetto (*Serenoa repens* Bartr.). A 1997 study postulated the mechanism for saw palmetto's and stinging nettle's action on BPH. For a decade now we have known that these two plants (plus the African prune, *Prunus africana* Hook) mimic the action of the popular prescription drug finasteride (Awang, 1997). Dennis Awang advances the hypothesis that components in stinging nettle, saw palmetto and African prune shift the androgen/estrogen ratios, involving estrogen and estradiol production from androstenedione, or, in other words, the same action as Wilson postulated.

A 1998 review article of studies on natural product drugs (crude extracts) pointed to seven of eight BPH studies that demonstrated "significant improvement in BPH symptoms in patients taking 320 mg of SPE (saw palmetto extract) for 1 to 3 months" (O'Hara et al., 1998). Dennis Awang attributes these herbs' "apparent superior potential" for BPH

---

[1] Department of History, Campus Box 8108, North Carolina State University, Raleigh, NC, 27695-8108.

Flavonoids in Cell Function, edited by Béla Buslig and John Manthey.
Kluwer Academic/Plenum Publishers, New York, 2002.

treatment to "the complexity of their clinical compositions affecting the multifactoral aspect of BPH pathogenesis, compared with unidimensional pharmaceuticals" (Awang, 1997). A. Colin Buck reports the advantages of these natural product drugs for treating BPC over synthetic α-adrenergic blockers and 5α-reductase inhibitors "would appear to be the lack of side-effects" (Buck, 1996).

In the first century of our era the Greek herbalist Dioscorides (*De materia medica* **4**:93) said that nettle was good for urinary problems, adding that the plant promotes menstruation, which was a common circumlocution for an abortifacient action. Finasteride, the synthetic drug, has contraindications: it is dangerous for pregnant women not only to take but even to handle finasteride or Proscar® tablets, and to receive blood transfusions or the semen of a finasteride user at a risk of potential abortion (PDR, 1993). Dioscorides said nothing about nettle promoting hair growth, but an even earlier writer, one of the authors in the Hippocratic Corpus said that it caused hair to grow (Hippocrates, 2. 189). More meaningful still, from Hippocrates to modern times, folk medicine employed nettle to promote the growth of hair. Grieve's Herbal reports that an "efficient Hair Tonic can be prepared from Nettle:

> Simmer a handful of young Nettles in a quart of water for 2 hours, strain and bottle when cold. Well saturate with lotion every other night. This prevents the hair from falling and renders it soft and glossy .... For stimulating hair growth, the old herbalists recommended combing the hair daily with express Nettle juice" (Grieve, 1971).

2. ST. JOHN'S WORT, MENTAL DEPRESSION, AND ANTIMICROBIAL ACTIONS

A recent issue of *The Lancet* reported that St. John's wort (*Hypericum perforatum* L.) has effective antibacterial action through its constituent compound, hyperforin (Schempp et al., 1999). The scientists who wrote the report began with the statement that St. John's wort is "a folk remedy commonly used for the treatment of skin injuries, burns, and neuralgia." In disclosing their laboratory data, the authors concluded that the folk remedy was the "rationale" for these treatments, and, by rationale, they meant scientifically established claims.

I do not know whether these scientists initiated their project because they wanted to test the rationality of folk usage or whether they selected this plant for other reasons and later learned of its use in traditional medicine. Had they researched thoroughly its folk usage, they would have learned that St. John's wort had been employed for thousands of years for just such treatments, long before its use for mental depression. Dioscorides employed the plant for bloody wounds in a plaster (Dioscorides 3. 156), but the ancients did not employ it for depression. The earliest use for St. John's wort as an antidepressant is in a prescription found in a Lorsch manuscript from around 800, when Charlemagne was the monastery's king. At Lorsch the plant was employed in a prescription, said to be from Galen (although this is unlikely to have been in his works), where it was given for *melancolia* (Stoll, 1992). Melancholy corresponds to the modern day diagnosis of mental depression.

An old English name for the plant was "touch-and-heal" (CRC, 2000). In this case, the name alone should have pointed the way to explore its pharmaceutical effects, even without historical research. Modern science should explore the history of a natural-product drug in addition to its chemistry and somatic effects. In doing so modern science's quest for new pharmaceuticals can be enhanced with historical investigation. In the example of St. John's wort, another commissioning of history is to learn of possible side effects.

Recently when I spoke at the First International Conference on St. John's wort (Anaheim, CA, USA, March 16-17, 1998), I suggested that, because of the persistence of the plant being named as an abortifacient, a prudent vendor should consider a label as a contraindication for pregnancy. Later, at the same conference, I learned from various speakers that chemical analyses reveal that rutin, a flavonoid that is a known abortifacient (Riddle, 1992; Duke, 1985), appears in some species and in varying amounts (ranging from a trace to quantities sufficient for pharmaceutical action) according to the phase of the development (Meier and Zeller, 1998, Grinenko, 1990, Makovets, 1999). The quantitative variation of rutin's presence is a good example of how difficult is to adapt natural-product drugs for modern pharmaceutical usage.

St. John's wort was employed in the ancient and medieval medicine for bladder problems, but modern laboratory testing has not examined the plant for possible kidney and bladder activities. Some studies note the claim in folk medicine (Denke et al., 1999; Duke, 1985), and a Russian study reported that continuously irrigation of the bladder after prostatic adenomectomy with an herbal infusion including St. John's wort prevented hemorrhagic and purulent inflammation with beneficial results (Davidov et al., 1995). The nature of ancient and medieval medical writings do not specify the action that the plant allegedly has on the bladder. Until it is scientifically investigated, it would be idle speculation whether the action was merely a diuretic or an important pharmaceutical activity. What should not be idle is a precaution about its use and a call for studies on urinary tract activity. At least one recent report described a woman who took St. John's wort and who was hospitalized with a serious bladder problem (Graedon and Graedon, 1999).

3. MASTIC FOR INTERNAL ULCERS

Mastic is a notable example of history indicating effective pharmaceutical uses that fell out of modern science's learning. Dioscorides, the Greek herbalist, said that mastic was an effective drug for various forms of internal bleeding. Mastic is a resin taken from the plant *Pistacia lentiscus* L., although, he said, effective drugs were found to come equally from the fruit, leaves, bark or root. Dioscorides said that mastic was good for the following:

1) bleeding exportations
2) shriveled (or wrinkled) intestines
3) dysenteric discharge of blood
4) internal bleeding
5) prolapsed uterus and anus
6) bleeding from uterus

Not all uses were for internal bleeding. Mastic also prevents *nomas* (a boil or ulcer?) from spreading, is diuretical, makes firm unstable teeth when washed with it, and its green sprigs are effective in cleaning teeth. The resin alone, when drunk, is good for bleeding exportations, old coughs, the stomach (but it causes belching), stimulating hair growth on eye brows, and is good in toothpaste because it cleans, makes white, strengthens, and gives good breathe (Dioscorides, 1. 70). Living about the same time, Pliny the Elder conveys much the same uses for mastic but added that it can be applied topically to cure abrasions. He related the anecdote that he knew himself how the physician Democrites cured Considia, daughter of the consul Marcus Servilius, by feeding her milk from a goat fed only on mastic (Pliny, 24. 28.42). Recently mastic was rediscovered to have very positive results when treating gastric and duodenal ulcers and demonstrated to have antimicrobal actions as well (Magiatis et al., 1999; Mansoor et al., 1986; Ali-Shtayeh and Abu Ghdeib, 1999).

Often the West makes its discoveries seemingly serendipitously or, less frequently, through insightful hunches. In the cardiovascular area, Dioscorides in the first century observed that fresh garlic "cleans the arteries"-- this is a precisely literal translation of the Greek *lamprunei arterias* (Dioscorides 2. 152). Now, almost 2,000 years later we rediscover (without knowing it was a rediscovery) that a compound in garlic, ajoene, and alliinase, an enzyme, inhibit the aggregation of blood platelets and exhibit fibrinolytic activities (Koscielny et al., 1999; Blumenthal, 2000).

4. SUMMARY

To trace the history of a natural product and its use, it is necessary to identify the correct plant among around a half-million species. One must also know how and when to harvest the plant and the morphology of location and extraction. Within the same species, plant chemistry varies, depending upon climatic and soil conditions, stage of maturity and even diurnal factors. To all of these variations must be added the diagnostic ability of physicians and native healers (to distinguish between Hippocratically-trained Western physicians and whose knowledge is less formally taught). Seldom was a disease identified as we know it today, but the constellations of symptoms described, when studied carefully within the framework historical setting of the culture, can be related to modern medicine. It is essential to study the historical contemporary usage data in the language in which those accounts were written. Translators are often philologists who are not sensitive to medical nuances. Modern readers of translated historical documents often are unaware of the precision the authors delivered in describing medical afflictions and their treatments.

Natural product drugs are truly products of human knowledge. Because so many modern pharmaceuticals are manufactured synthetically we forget that once either the compound or its affinity had a home in a natural product. Over 2,500 years ago man first used a drug obtained from white willow bark, which was aspirin or acetylsalicylic acid. Today's scientists continue to be bewildered by just what aspirin's mechanisms of actions are, discovering new modes of action, and how they relate to medical diagnostics. Whatever the science of aspirin, an intelligent person today takes it just as our ancestors did for millennia. Throughout time, explanations continue to vary just as purpose of admini- stration do as well. Nevertheless, aspirin is perceived as being beneficial.

Historical in-use data can also be a factor in judging a drug's safety, since the records of its use provide observations made by intelligent persons over generations of employment. Many historical "drugs" have crossed the line from drug to food. A number of them are now common items on our tables: coffee, tea, sugar, lemon, chocolate, pepper, to name a few. The example of coffee affords useful insight. It was first employed as a drug (like tobacco), its botany and chemistry are well known, it has been in widespread use for centuries with diverse ethnic populations in a variety of preparations and amounts consumed. Still we are unsure about coffee's effect on health, the latest assertion being that the caffeine it contains may delay the onset of Alzheimer's. In contrast, the mercury drugs were in widespread use for a long period of time by many populations and that fact indicates that the toxic tolerance in humans is probably higher than as currently proscribed.

The past contains important data for the scientific investigator. Like any field of research, historical investigation requires specialized knowledge, but much of that knowledge is readily accessible and employable. Rediscovery through examination of historical contemporaneous use data can be efficient and relatively easy compared to the travails of original research in pursuit of a discovery only to learn later that our ancestors had already made that discovery through trial and error in human usage.

If we had started a search from the clues provided by history, presumably our discoveries would have been earlier, and we would have benefitted. As it is, we learn history but not science or else we learn science, not history. Both taken together the learning can be enhanced.

5. REFERENCES

Ali-Shtayeh, M. S., and Abu Ghdeib, S. I., 1998, Antifungal activity of plant extracts against dermatophytes, *Mycoses* **42**:665-672.

Awang, D., 1997, Saw palmetto, african prune and stinging nettle for benign prostatic hyperplasia (BPH), *Can. Pharm. J.* **130**:37-40, 43-44, 62.

Blumenthal, Mark (ed.), 2000, *Herbal Medicine*. Expanded Commission E Monographs. American Botanical Society, Austin, TX, pp. 139-48.

Buck, A. C., 1996, Phytotherapy for the prostate, *Brit. J. Urol.* **78**:325-336.

CRC Publishing Co., 2000, *CRC World Dictionary of Plant Names*, 4 vols. CRC Press, Boca Raton, **2**:1286.

Davidov, M. I., Goryyunov, V. G., Palagin, P. M., Lyadov, A. A., Kharichev, S. V., Petrunyaeu, A. L., Kibanov, V. P., Shilov, A. P., Shkraraburov, G. P., and Veretennikov, V. B., 1995, Postadenomectomy phytoperfusion of the bladder [in Russian], *Urologiya i Nefrologiya*. **0/5**:19-20 [through Biological Abstracts].

Denke, A., Schempp, H., Mann, E., Schneider, W., and Elstner, E. F., 1999, Biochemical activities of extracts from *Hypericurn perforaturn* L. 4th Comm.: Influence of different cultivation methods, *Arzneimittelforschung* **49**:120-125.

Dioscorides, 1958, *De Materia Medica*, Wellmann, M., ed., in 3 vols., Weidmann, Berlin

Duke, J.A., 1985, *CRC Handbook of Medicinal Herbs*, CRC Press, Boca Raton, FL, p. 243, p.417.

Graedon, Joe and Graedon, Terry, 1999, People's Pharmacy, Raleigh News and Observer, Nov. 7, p. 6D.

Grinenko, N. A.,1990, Composition of flavonoids and anthraquinone derivatives in *Hypericum perforatum* L. and *Hypericum maculatum* Crantz., [in Russian] *Rastitel'nye Resursy* **25/3**:387-392 [through Biological Abstracts].

Grieve, M., 1971, A Modem Herbal, 2 vols., Dover, New York, **2**:578.

Hippocrates, 1973, *Gynaikeon*, E. Littré, ed. Hakkert, Amsterdam, **8**:370.

Horton R., 1992, Editorial: Benign prostatic hyperplasia: new insights, *J. Clin. Endocrinol. Metab.* **79**:1115-1121.

Koscielny, J., Klüßendorf, D., Latza, R., Schmitt, R., Radtke, H., Siegel, G., and Kiesewetter, H., 1999, The anti-atheroschlerotic effect of *Allium sativum*, *Atheroschler.* **144**:237-249.

Magiatis, P., Melliou, E., Skaltsounis, A. L., Chinou, E. B., and Mitaku, S., 1999, Chemical composition and antimicrobal activity of the essential oils of *Pistacia lentiscus* var. chia, *Planta Med.* **65**:749-752.

Makovets, K., 1999, Research of biologically active substances of *Hypericum* L. species [in Ukranian]. *Farm. Z.*, **0/5**:38-44.

Mansoor, S., Al-Said, M. S., Ageel, A. M., and Parmar, N. S. 1986, Evaluation of mastic, a crude drug obtained from *Pistacia lentiscus* for gastric duodenal antiulcer activity, *J. Ethnopharmacol.* **1513**: 271-278.

Meier, B., and Zeller, A. G., 1998, Analytical techniques to assess quality and dosage requirements for *Hypericum perforatum* L. Proceedings St. John's Wort Symposium, American Herbal Products Association, Anaheim, CA, p. 5 et passim. in Proceedings.

O'Hara, M., Kiefer D., Karrell K., and Kemper, K., A review of 12 commonly used medicinal herbs, *Arch. Fam. Med.* 7:534.

PDR, 1997, *Physicians' Desk Reference*, 47th ed., Medical Economics Publishing, p. 1598.

Pliny, 1980, *Natural History.* 10 vols. Jones, W. H. S., ed., and trans., Harvard University Press, Cambridge, MA.

Riddle, J. M., and Estes, J. W., 1992, Oral contraceptives in ancient and medieval times, *American Scientist* **80**(3): 226-234.

Schempp, C. M., Pelz, K., Wittmer, A., Schöpf, E., and Simon, J. C. 1999, Antibacterial activity of hyperforin from St. John's wort, against multiresistant *Staphylococcus aureus* and gram-positive bacteria, *Lancet* **353**:2129.

Stoll, U., 1992, *Das Lorscher Arzneibuch*. Franz Steiner, Stuttgart, from Lorsch MS Bamberg Med. 1:97, fol. 49.

Wilson, J. D., 1972, Recent studies on the mechanism of action of testosterone, *New Engl. J. Med.* **287**, 1284-1291.

POTENTIAL HEALTH BENEFITS FROM THE FLAVONOIDS IN GRAPE PRODUCTS ON VASCULAR DISEASE

John D. Folts[1]

1. INTRODUCTION

In spite of recent advances in diagnosis and treatment, coronary artery disease and cerebral artery disease continue to be the leading causes of mortality and morbidity in the United States (Scharf and Harker, 1987; Lenfant, 1999). More than 1 million people had heart attacks last year in the U.S. and one-third of them died from their first heart attack (Lenfant, 1999). Thus there is considerable interest in the prevention of coronary and cerebral artery disease. The underlying pathogenesis of these problems is atherosclerotic vascular disease. This is marked by the chronic silent evolution of atheromas over many years, from as early as age 10 up to ages 40-50 years. Then an acute thrombotic phase appears with plaque rupture, the onset of unstable angina, and fatal or nonfatal myocardial or cerebral infarction (Scharf and Harker, 1987; Loscalzo, 1990; Ross, 1986).

Atherosclerosis is a multi factorial disease, involving risk factors which have been known for many years. Cigarette smoking, diabetes, elevated LDL cholesterol, high blood pressure, age, male sex, elevated Lp(a) and homocysteinemia, are some of the key ones (Rabbani and Loscalzo, 1994)

While the etiology of atherosclerosis is very complicated, the interactions among four cell systems are thought to be involved in atherosclerotic development and the acute occlusive "clot that kills." They are endothelial cells, platelets, vascular smooth muscle cells and monocytes/macrophages (Rabbani and Loscalzo, 1994). They will be described briefly and the potential benefits of antiplatelet/antioxidant therapy with the flavonoids in red wine and purple grape juice will be presented.

---

[1] Address for Correspondence: John D. Folts, Ph.D., F.A.C.C., Director, Coronary Thrombosis Research Laboratory, Professor of Medicine, Cardiology, University of Wisconsin Medical School, 600 Highland Ave., H6/379, Madison, WI 53792-3248. Telephone: (608) 263-1543, FAX: (608) 263-0405. E-mail: jdf@medicine.wisc.edu

Flavonoids in Cell Function, edited by Béla Buslig and John Manthey.
Kluwer Academic/Plenum Publishers, New York, 2002.

1.1. Endothelial Cells, Endothelium

The most important function of normal, healthy vascular endothelium, consisting of a layer of endothelial cells (ECs) is to provide a thromboresistant surface which prevents the adhesion of blood cells such as platelets and white blood cells to the arterial wall (Vita et al., 1996). A typical artery is shown in cross section and longitudinal section in Fig. 1.

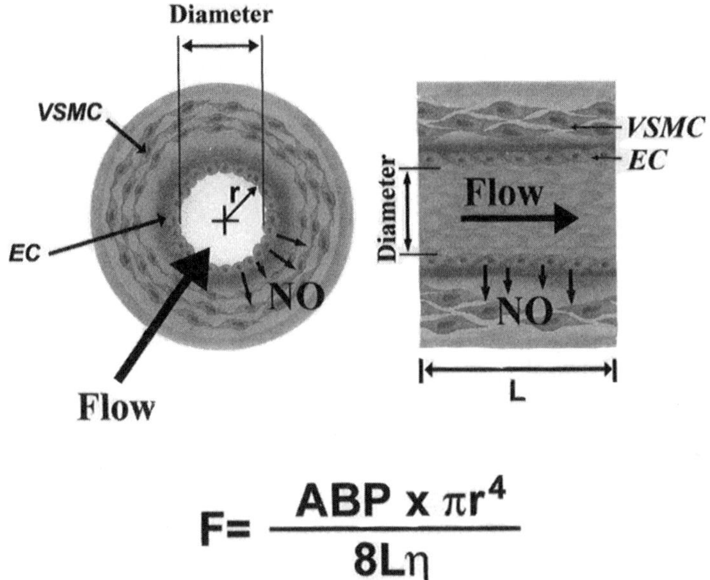

$$F= \frac{ABP \times \pi r^4}{8L\eta}$$

Figure 1. Nitric oxide's role in flow-mediated arterial dilation. The schematic of an artery is shown in cross section on the left and in longitudinal section on the right. Vascular smooth muscle cells (VSMC) are shown circumferentially located on the left, along with the endothelial cells (EC) at the luminal surface. Cholinergic nerves normally innervate arteries and can release acetylcholine (ACH) which stimulates the ECs to produce more nitric oxide (NO). This in turn acts on the VSMC and relaxes them thereby producing vasodilation. This results in an increase in radius (r) that can permit a large increase in flow, as blood flow (F) is proportional to the arterial blood pressure (ABP), and the radius (r) to the fourth power. The other factors determining flow through the artery are the length (L) of the vessel and the viscosity ( ) of the blood. If blood flow is increased due to downstream arteriolar dilation, the increased flow produces increased shear forces on the endothelial cells. This increased shear stimulates the Ecs to make more NO which also causes the artery to dilate.

The endothelial cells (EC) can be seen lining the inner wall of the artery (Fig. 1). The endothelial cell layer also acts as a semipermeable membrane which inhibits substances such as cholesterol from entering the arterial wall (Vita et al., 1996). The ability of the endothelium to provide these and other protective functions can be markedly reduced by endothelial injury. This injury produces endothelial dysfunction and is thought to be one of the earliest events in the development of the atherosclerotic process (Vita et al., 1996). Endothelial cells also cover established atherosclerotic plaques, and if relatively healthy, protect against platelet mediated occlusive thrombosis (Vita et al., 1996).

When blood flow increases through a coronary or other artery this increased blood flow passing over the endothelial cells stimulates them to secrete extra amounts of endothelial derived relaxing factor now thought to be nitric oxide (NO) (Rubanyi et al., 1986). This increased NO stimulates the underlying vascular smooth muscle cells (VSMCs) to relax and this in turn dilates the artery and allows for even more blood flow (Fig. 1) (Cooke et al., 1990). A small increase in diameter or radius can make a large difference in flow since in the formula for flow (F), the radius (r) is taken to the fourth power. The nitric oxide also inhibits platelet and monocyte adhesion on the arterial wall (Fig. 1, Fig. 2)

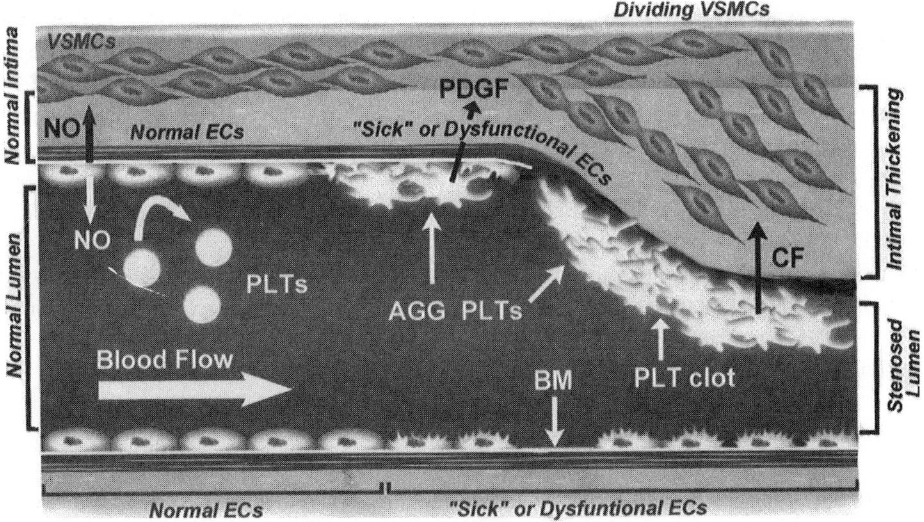

Figure 2. Platelet-mediated intimal thickening. This is a schematic figure of an artery depicting normal endothelial cells (EC) on the left, and "sick" or dysfunctional ECs in the center and on the right. Healthy ECs secrete nitric oxide (NO) which does several useful things. It inhibits platelets (PLT) and other white blood cells from adhering to the arterial wall. The released NO also diffuses to the vascular smooth muscle cell (VSMC) and relaxes them. In addition, normal amounts of NO can inhibit the division and migration of VSMCs toward the lumen. When the ECs are sick or dysfunctional, or when they have sloughed off the underlying basement membrane (BM), they do not make enough NO to inhibit platelet adhesion and aggregation (AGG) platelet aggregates form on the wall. Platelets (and other cells) release platelet-derived growth factors (PDGF) which stimulate VSMCs to divide. The platelets also release chemoactive factors (CF) which draw the dividing platelets down toward the lumen. This produces intimal thickening, which results in a stenosed lumen.

Endothelial cells can be damaged or made dysfunctional by hypertension, increased LDL cholesterol, diabetes, cigarette smoking, free radicals, or elevated plasma homocysteine (Vita et al., 1996). Endothelial production of NO is known to be impaired in atherosclerosis (Bossaller et al., 1987). Finally, there is evidence that loss of endothelium derived NO contributes to acute coronary occlusive events and heart attacks (Okumura et al., 1992; Bogaty et al., 1994). Theoretically, any drug or dietary substance that would help "heal" the endothelium, allowing it to make more NO should help reduce the development and progression of cardiovascular disease.

1.2. Vascular Smooth Muscle Cells (VSMCs)

Vascular smooth muscle cells (VSMCs) are found in a circumferential path in the arterial wall in the cross section shown on the left of Fig. 1. Their primary function is to contract or relax, thus regulating the diameter of the artery (Fig. 1, Fig. 2). They are normally stimulated to relax, (permitting vasodilation) by several substances including nitric oxide (NO) or a form of NO, which is released by healthy ECs (Rubanyi et al., 1986; Cooke et al., 1990). This NO also inhibits pathological changes in the behavior of the VSMCs.

One of the key features of the early atherosclerotic development process is a change in the behavior of these VSMCs. They can be stimulated to divide and multiply by mitogens or growth factors released by activated platelets and other cells (shown on the upper right of Fig. 2) (Rabbani and Loscalzo, 1994; Navab et al., 1996). If there is damage to the endothelium, i.e., if the ECs are sick or dysfunctional, there will be less NO produced and then platelets and other cells can stick on the arterial wall (Fig. 2). They release platelet derived growth factor that cause the VSMCs to start multiplying as shown on the upper right of Fig. 2 (Rabbani and Loscalzo, 1994). In addition, the dividing VSMCs also begin to move toward the platelets which are secreting the growth factors. This migration of VSMCs produces intimal thickening and causes narrowing of the arterial lumen (Fig. 2) (Rabbani and Loscalzo, 1994; Navab et al., 1996). Substances that inhibit the pathological activity of the platelets and reverse the damage to the ECs, could reduce the abnormal behavior of the VSMCs and the rate at which one gets atherosclerotic narrowing of the coronary or other important arteries.

1.3. Platelets

Platelets are small cells that circulate in the blood. Their main function is to stick to small tears in tiny arteries and capillaries that are damaged in the normal course of living. They prevent bleeding and promote healing in this area of damage. They were not intended to play a major role when there is damage to the endothelium of large arteries such as the coronary arteries. They do not adhere or stick to normal healthy endothelium because the ECs release substances such as NO which inhibit platelets (PLTs) from sticking (Fig. 2) (Rabbani and Loscalzo, 1994). However, if the endothelial cells are damaged or sloughed off, platelets can now stick and aggregate (AGG) on the arterial wall (Fig. 2).

Activated platelets then secrete the growth factors that cause smooth muscle cells to divide and also to migrate down into the lumen causing intimal thickening (Fig. 2). Platelet inhibitors turn down this platelet-mediated phenomenon and thus should reduce this contribution by overactive platelets to the atherosclerotic process and also platelet-mediated acute occlusive thrombosis.

1.4. Monocyte/macrophages And Free Radicals

When the endothelium is damaged or dysfunctional it is easier for LDL cholesterol to diffuse into an arterial wall (Navab et al., 1996). LDL cholesterol can be oxidized by free radicals generated within the arterial wall (Fig. 3). There are a variety of these free radicals formed in the body as well as those generated by cigarette smoking and air pollution. All the cells in the artery wall, endothelial cells, smooth muscle cells, monocytes and macrophages are able to produce free radicals which can oxidize and modify LDL cholesterol (Navab et al., 1996; van Hinsbergh et al., 1986). When LDL is modified by free radicals it may become trapped in the arterial wall. Oxidized or modified LDL (M-LDL), may then contribute to a series of deleterious reactions (Fig. 3). The M-LDL stimulates ECs to release cellular adhesion molecules (CAMs) which attract monocytes and other white blood cells (Fig. 3). Monocyte chemotactic protein (MCP) is also secreted which draws monocytes from the flowing blood into the arterial wall. The monocyte then gets converted to a macrophage (MAC) (Navab et al., 1996) (Fig. 3).

Macrophages are renegade cells with unusual properties. Most cells in the body, including macrophages, take up native LDL cholesterol at a regulated or controlled rate. This means that after the cell takes up some cholesterol, the LDL receptors are down regulated and it does not continue to take up more. This protects the cell from excess accumulation of cholesterol. The macrophage (MAC) also can down regulate its LDL receptors if it takes in native or unmodified LDL. However, the macrophage also possesses scavenger receptors which can take up modified or oxidized LDL (M-LDL) in an unregulated fashion. Thus, they continue to take up cholesterol until it becomes engorged with LDL-C and now becomes a "foam cell" (Fig. 3) (Navab et al., 1996). Excess accumulation of foam cells contributes to the fatty streak in the arterial wall and is thought to be a precursor to the atherosclerotic plaque (Navab et al., 1996; van Hinsbergh et al., 1986; Holvoet and Collen, 1998). The atherosclerotic mechanisms drawn in Figure 2 and 3 are illustrated separately

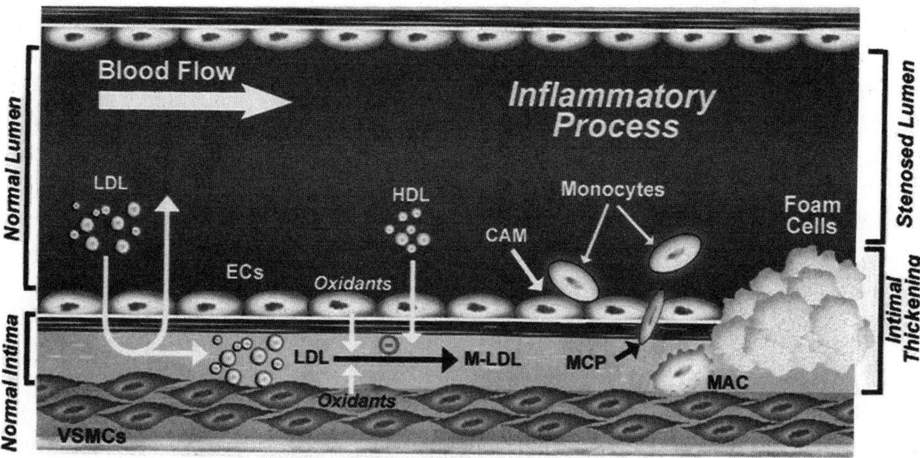

Figure 3. Oxidation of LDL leads to the formation of foam cells. Low density lipoprotein (LDL) are known to enter the arterial wall and contribute to the development of foam cells and the fatty streak, the first visible sign of atherosclerotic development. If LDL becomes oxidized by oxidants or free radicals released from endothelial cells (ECs), vascular smooth muscle cells (VSMCs) and other cells, it become modified LDL (M-LDL). Modified LDL can stimulate ECs to release cellular adhesion molecules (CAM) which attract monocytes and T lymphocytes to attach to the arterial wall. This is more likely to happen if the ECs are sick or dysfunctional. M-LDL can also stimulate the production of monocyte chemotactic protein (MCP) which draws monocytes into the intimal space. Then M-LDL can convert monocytes into macrophages (MAC) scavenger cells. The MACs take up M-LDL in an unregulated fashion and become engorged with cholesterol and become foam cells. When a group of these foam cells accumulate on the arterial wall, they appear as a fatty streak, the first visible sign of atherosclerosis.

for clarity. Actually, these mechanisms can occur at the same time, in the same place. Thus, when platelets aggregate on the arterial wall they release free radicals which can oxidize adjacent LDL and damage endothelial cells (Aviram et al., 1990). Conversely, if resting platelets come in contact with oxidized LDL this causes the platelets to aggregate (Rabbani and Loscalzo, 1994). So, there is a complex deleterious interaction between platelets, free radicals, LDL, and ECs occurring simultaneously in the developing atherosclerotic plaque (Rabbani and Loscalzo, 1994). In addition, both free radicals and oxidized LDL can damage endothelial cells, impair their release of NO, and make it easier for platelets to adhere and aggregate on the arterial wall.

The presence of increased amounts of antioxidants in the blood may help protect against the oxidation, or modification of LDL (cholesterol), by free radicals released by platelets and other cells, leading to the deleterious processes described above. Also, if these antioxidant substances helped to improve endothelial function, the LDL would be less likely to pass through the endothelium and into the arterial wall where it causes damage, and blood cells would be less likely to attach themselves to the arterial wall. Thus it would seem beneficial to have something in the diet which had platelet inhibiting properties, improved endothelial cell function, and protected LDL cholesterol from oxidation.

2. DOES MODERATE ALCOHOLIC BEVERAGE CONSUMPTION DECREASE THE RISK OF HEART ATTACKS?

It has been suggested that the moderate consumption of alcoholic beverages reduces the incidence of fatal and nonfatal heart attacks (Kauhanen et al., 1999; Maclure, 1993). This is often attributed to the modest increase in HDL that occurs with moderate alcoholic beverage consumption, and is usually attributed to the alcohol content (Maclure, 1993). However, we and others have shown that red wine, dealcoholized red wine, and purple grape juice appear to have three properties unrelated to alcohol content that may reduce the rate of development of atherosclerotic narrowing of coronary and other arteries. These properties are antiplatelet, antioxidant and improvement in endothelial cell function.

3. FLAVONOIDS ARE NATURAL ANTIPLATELET/ANTIOXIDANT SUB-STANCES FOUND IN THE DIET

Flavonoids are polyphenolic compounds widely distributed in the plant kingdom but are not found in any animal food sources (Cook and Samman, 1996). Plant preparations containing flavonoids have been used for centuries as herbal remedies for a variety of diseases including allergy, arthritis, and cancer (Cook and Samman, 1996). The pioneering work of Szent-Györgyi demonstrated the capacity of *Capsicum annuum* and *Citrus limon* fruits to cure subcutaneous capillary bleeding (Rusznyák and Szent-Györgyi, 1936).

There are a remarkable number of physiological and pharmacological effects of these plant extracts (Cook and Samman, 1996, Gryglewski, 1987). Many of them are found in grape products and have antioxidant, anti- inflammatory, and platelet inhibitory effects (Cook and Samman, 1996, Gryglewski, 1987). There are many polyphenolic compounds including the flavonoids found in grapes, such as catechin, epicatechins, kaempferol, anthocyanins, procyanidins, and multimers of these individual monomers (Singleton, 1981; Ribereau-Gayon, 1982; Silva et al., 1991; Vinson, 1998).

4. ANTIPLATELET PROPERTIES OF RED WINE AND PURPLE GRAPE JUICE

4.1. Animal Studies

Red wine and grape products have been shown to have both antioxidant and antiplatelet properties *in vitro* and in animals and man. (Cook and Samman, 1996, Gryglewski, 1987, Demrow et al., 1995; Frankel et al., 1993; Fuhrman et al., 1995; Folts, 1998). We recently demonstrated under controlled conditions that both red wine and purple grape juice have significant antiplatelet properties *in vivo* in animals and *ex vivo* in humans. Red wine (4 ml/kg) and purple grape juice (10 ml/kg), given orally, in our well-established and characterized canine model of experimental coronary artery stenosis and thrombosis (Folts cyclic flow model) significantly inhibited platelet activity, periodic acute platelet-mediated coronary thrombosis, and the associated cyclic flow reductions (CFRs) (Demrow et al., 1995). This was confirmed by *ex vivo* whole blood platelet aggregation studies (Demrow et al., 1995). The amount of purple grape juice given as a single dose to the anesthetized dog was 10 ml/kg or the equivalent of approximately 20 oz of juice in an adult (80 kg) human. In these studies when platelet activity had been significantly inhibited by purple grape juice, the platelet activity was not renewed by increasing the plasma epinephrine levels (Demrow et al., 1995). This animal model of stenosis, with intimal damage has been described as a good model of human unstable angina (Sherry, 1984). We have previously shown that if experimental coronary thrombosis is abolished and platelet activity significantly reduced with aspirin, raising the plasma epinephrine levels renew platelet activity about 60% of the time (Folts and Rowe, 1988; Folts et al., 1999). Thus in the animal model purple grape juice is a better platelet inhibitor than aspirin.

We also conducted a study of grape juice in cynomologous monkeys (Osman et al., 1998). Five monkeys had a control blood sample drawn for *ex vivo* whole blood platelet aggregation studies. Platelet activity was determined with a whole blood platelet aggrego-meter (Demrow et al., 1995; Osman et al., 1998). This device is often used to test the effect of oral platelet inhibitors on human blood (Cardinal and Flower, 1980). A blood sample was drawn before initiating the juice feeding study and studied in the aggregometer. The monkey was then fed half the dose required when given as a single dose (5 ml/kg) of purple grape juice for 7 days. Then a second blood sample was drawn for platelet aggregation studies. By comparing the platelet aggregation responses after 7 days of feeding purple grape juice with the control sample obtained before the grape juice was given, we can determine how much the platelet activity has been turned down by consuming the juice. Feeding the monkeys the purple grape juice inhibited platelet activity induced by collagen as an agonist by 41%, (p<.01) (Osman et al., 1998). In addition, the purple grape juice blocked the synergism between collagen and epinephrine induced platelet aggregation, whereas aspirin does not (Osman et al., 1998). Finally, when these monkeys were fed equal amounts of orange juice or grapefruit juice for 7 days there was no change in platelet activity (Osman et al., 1998). Thus, the specific flavonoids in purple grape juice appear to be more potent at inhibiting platelets than those in orange or grapefruit juice.

4.2. Human Studies With a Single Dose of Purple Grape Juice and Red Wine: Antiplatelet Effects

Ten healthy human subjects, 5 men and 5 women aged 21-55 years, were chosen from the university community. They abstained from drinking alcoholic beverages, grape or other fruit juices, and tea for one week prior to the study and throughout the study. They also abstained from all medications, including aspirin products (Folts, 1998). Since we wished to study the effects of flavonoids found in fruits and vegetables, no vegetarians or people consuming 3 or more servings of fruits and vegetables per day were utilized in these studies.

The subjects had a blood sample drawn for whole blood platelet aggregation studies and then they consumed either a single dose of 5 ml/kg of red wine (Chateau Neuf du Pape 1998), white wine (Chateau Villotte Bordeaux 1990) or 12 ml/kg of Welch's purple grape juice in a randomized fashion, with a week of washout between each beverage. Blood was drawn 2 hours after the consumption of a single dose of each beverage, and platelet aggregation studies were done. Consuming red wine inhibited platelet activity by 39 ± 11% (p<.01) whereas an equal amount of white wine did not significantly inhibit platelet activity (Folts, 1998). Consumption of purple grape juice by the 10 human subjects also inhibited platelet activity by 34 ± 10% (p<.03). There are 7-10 times more total flavonoids in red wine and purple grape juice compared to white wine (Waterhouse et al., 1997; Waterhouse and Teissedre, 1997). This may be why these doses of red wine and purple grape juice consumption significantly inhibited *ex vivo* human platelet activity but the white wine did not (Folts, 1998)

4.3. Human Study of the Antiplatelet Effects of 7 Days of Feeding Purple Grape Juice

Flavonoids are thought to bind to tissues and proteins. Thus we tested half the single dose of purple grape juice in 10 more healthy subjects, (5 males, 5 females) ranging in age from 26 to 58 years (Keevil et al., 2000). In order to minimize the effects of diet, drugs, and general health on platelet function, the same dietary restrictions described above were imposed. Each subject drank 5-7 ml/kg of purple grape juice for 1 week with a one-week washout prior to the study. The subjects kept a daily diary of food and juice consumption to confirm compliance.

After a control blood sample was drawn for whole blood platelet aggregation studies the subject consumed 5-7 ml/kg of purple grape juice for 7 days. A second blood sample was then drawn for repeat platelet aggregation studies. The 7 day consumption of purple grape juice inhibited platelet activity by 77% (p<.003) (Keevil et al., 2000). Although the inhibition of platelet activity was quite significant, there was considerable variation in the standard deviation of the responses (± 13%) of the 10 subjects to the inhibitory effects of the purple grape juice. This may be due to the normal biological variation in platelet activity between men and women, variations in diet, and basal catecholamine levels. In this study, after a 7-day washout period, the subjects were also studied before and 7 days after consuming an equal amount of orange juice or grapefruit juice for 7 days. There was no measurable change in platelet activity from these two citrus juices (Keevil et al., 2000). The results of this small study of the inhibitory effects of purple grape juice on platelet activity in human volunteers are supported by our animal studies (Demrow et al., 1995, Osman et al., 1998).

5. HUMAN STUDIES WITH PURPLE GRAPE JUICE AND RED WINE: ENDO-THELIAL EFFECTS

5.1. Coronary Risk Factors and Endothelial Dysfunction in Humans

Vascular endothelium plays an important role in the regulation of vasomotor tone in coronary, brachial and other arteries (Vita et al., 1996; Solzbach et al., 1997). In normal arteries when flow increases the increased shear forces stimulate the ECs to make more NO. This in turn relaxes the VSMC and increases the arterial diameter (Fig. 1). Abnormalities in endothelial mediated vasomotor dilation responses have been observed in patients with atherosclerosis, hypercholesterolemia, diabetes mellitus, cigarette smoking, hyperhomo-cysteinemia, and essential hypertension (Vita et al., 1996; Solzbach et al., 1997). This impairment in endothelial mediated vasodilation has been linked to the reduced availability of NO from endothelial cells (Vita et al., 1996; Ting et al., 1997). It is thought that there is either impaired endothelial NO production, or enhanced inactivation of NO by vascular superoxide or other free radicals (Ting et al., 1997). Antioxidants may neutralize these free radicals, thus helping to preserve NO levels, and therefore maintain endothelial function.

5.2. Effects of Red Wine and Purple Grape Juice on Endothelial Function

A series of experiments was done on isolated arterial rings in a tissue bath by Fitzpatrick et al. which showed that the flavonoids in purple grape juice and red wine could increase the production of NO by endothelial cells thus causing relaxation of the adjacent vascular smooth muscle cells (Fitzpatrick et al., 1993). First, the ring is precontracted by phenylephrine (PE). Then a certain concentration of acetylcholine (ACH) which stimulates ECs to make more NO was then added to the tissue bath and the arterial ring relaxed. This was due to the stimulation of the endothelial cells nitric oxide synthase (NOS) by ACH, to produce increased amounts of NO. The PE and ACH can be washed out of the tissue bath, thus allowing further studies on the same arterial ring with new substances added to the bath. The ring can be contracted again with the same amount of PE. When a dilute solution of purple grape juice was added to the tissue bath, and the same amount of ACH added, the ring relaxed to a greater extent than previously. In addition, if L-arginine, the precursor of NO, was added to the bath the response was enhanced even more. Finally, if L-NNMA, a NOS inhibitor, was added to the bath the effect of the purple grape juice was blocked. This suggested that the purple grape juice enhanced the production of NO by the endothelial cells when they were given the ACH (Fitzpatrick et al., 1993). The increased NO produced by the grape juice enhanced relaxation of the vascular smooth muscle producing more vasodilation. Very similar results were obtained with red but not white wine or pure alcohol (Fitzpatrick et al., 1993). These studies provide a link to the flavonoids in purple grape juice and improvements in endothelial function. The results of these animal studies led to the following studies with grape juice in patients with coronary artery disease. When blood flow is increased in a healthy human coronary or brachial artery, the increased flow over the endothelial cells stimulates them to release increased amounts of NO (Rubanyi et al., 1986). This NO then diffuses to the adjacent VSMCs and stimulates them to relax (Fig. 1, Fig. 2). Studies have also been done in patients with coronary or other arterial vascular disease, showing that they have impaired endothelial function in their coronary and also in their brachial arteries (Takase et al., 1998,Schroeder et al., 1999).

5.3. Study of Endothelial Function and Flow Mediated Vasodilation in Patients with Stable Coronary Artery Disease: Effect of Oral Purple Grape Juice and Red Wine

We studied 15 patients with stable coronary artery disease before and again after they had consumed 5-7 ml/kg of purple grape juice daily for 14 days (Stein et al., 1999). Resting blood flow and brachial artery diameter were measured noninvasively by Doppler echo techniques (Stein et al., 1999). A large increase in brachial artery blood flow was then induced by first inflating a pneumatic blood pressure tourniquet placed around the widest part of the forearm to a pressure of 50 mmHg greater than the systolic blood pressure for the patient. This temporarily cuts off blood flow to the forearm, and causes transient ischemia. When the pressure cuff is suddenly deflated there is a large increase of blood flow through the brachial artery which supplies blood to the forearm. This increase in brachial artery blood flow is called a hyperemic response, and as the blood rushes past the endothelial cells in the brachial artery the increased flow and increased shear forces stimulate the ECs to secrete more NO (Fig. 1).

If the vasculature and the ECs were normal, i.e. no endothelial dysfunction, the measured brachial artery diameter as measured by the ultrasound probe would increase, i.e. the artery would undergo flow mediated dilation of up to 10-15%. If there is some endothelial dysfunction, as there is in patients with CAD, the increase in measured brachial artery diameter might be only 2-5% or less with the hyperemic response (Takase et al., 1998). In our 15 patients in the control state, there was very little increase in brachial artery diameter, with the induced hyperemia. This indicates that the patients have severe endothelial dysfunction. The average for the 15 patients was just 2.2% ± 2.9%, with 2 patients below the zero line, indicating that they were actually vasoconstrictors and had very serious endothelial dysfunction. However, after 14 days of consuming 5-7 ml/kg of purple grape juice daily, they significantly improved their endothelial function and flow-mediated dilation (FMD) such that now they averaged a 6.4% ± 4.7% (p <0.001) increase in brachial artery diameter. This is a significant improvement in FMD and endothelial function after just 14 days of grape juice consumption (Stein et al., 1999). An improvement in endothelial function may reduce the contributions of endothelial dysfunction to the atherosclerotic process described above.

While we studied the effects of purple grape juice on endothelial function in the brachial artery, it has been shown that there is a close relationship between endothelial function in the human coronary artery compared to the peripheral arteries (Anderson et al., 1995). Improvement in flow-mediated arterial dilation (and presumably endothelial function) has been shown in patients with administration of HMG-CoA reductase inhibitors (also called statins) and also vitamin E (Neunteufl et al., 1998,Wagner et al., 2000)

It is interesting to note that the 15 patients in our study had significantly impaired FMD (2.2%) in spite of the fact that most patients were taking both vitamin E (400 IU/day) and statins to lower their lipids. In addition, they were able to significantly raise their FMD to 6.6% by drinking the grape juice (Stein et al., 1999). Another grape beverage, red wine, consumed acutely also increased brachial artery FMD measured in a very similar fashion to our study in 7 healthy men (Itakura, 1999).

6. ANTIOXIDANT PROPERTIES: RED WINE AND PURPLE GRAPE JUICE

Antioxidant supplementation has become a widespread phenomenon in our society. Antioxidant vitamins and other nutraceutical supplements are widely consumed although the scientific evidence for this practice has not been firmly established (Duthie and Bellizzi, 1999). Both healthy physicians and nutritionists, as well as the lay public, are consuming large amounts of Vitamin C and E daily, although there is no formal recommendation for this consumption. A recent paper, *Antioxidant Vitamins and the Prevention of Coronary Heart Disease*, by Adams et al., (1999). recommends that healthy individuals should take supplements of vitamins C and E, along with 5-7 servings of fruits and vegetables daily, and make aggressive changes toward a healthy lifestyle to reduce the risk of coronary artery disease (Adams et al., 1999). Many of the flavonoids in grapes have antioxidant properties *in vitro* (Cook and Samman, 1996,Vinson, 1998,Waterhouse et al., 1997). A recent study examined the total antioxidant capacity of several commercial juices: purple grape, grape-fruit, orange, tomato and apple. The grape juice had almost four times the antioxidant capacity of all the other juices (Wang et al., 1996). Many flavonoids in fruits and vegetables have been shown to have strong antioxidant properties *in vitro* , often more potent than vitamin E. (Cook and Samman, 1996, DeWhalley, 1990, Pignol et al., 1988, Fraga et al., 1987, Cavallini et al., 1978, Afanas'ev et al., 1989, Das and Ratty, 1986, Kinsella et al., 1993, Facing et al., 1994). The flavonoids in red wine have been shown to inhibit copper-induced oxidation of human low density lipoprotein better than vitamin E (Frankel et al., 1993). However, one must keep in mind that there are a variety of free radicals generated in the body and a given antioxidant may neutralize only certain free radicals (Duthie and Bellizzi, 1999). Thus, we must not be misled into thinking that a given antioxidant can protect against all free radicals.

The protective antioxidant effect of red wine is thought to be due to phenolic compounds including procyanidins which have been shown to be potent free radical scavengers *in vitro* (Frankel et al., 1993). Their direct action on peroxyl radicals, formed by oxidative attack on lipid membranes and lipoproteins, may be a potential mechanism for their protection of LDL *in vivo* (Sies, 1993, Esterbauer et al., 1992). There are several *in vitro* studies demonstrating protection of LDL by red wine, supporting this hypothesis (DeWhalley, 1990, Yuting et al., 1990). Components of red wine and red grape juice, such as anthocyanins, procyanidins, and other flavonoids, can enrich LDL-C and make it less susceptible to *ex vivo* oxidation (Vinson et al., 1995)

6.1. Human Studies with Purple Grape Juice and Wine: Antioxidant Effects

Data showing *in vivo* antioxidant properties of red wine are limited. Red wine, 5 ml/kg body weight was given to healthy volunteers, and blood was sampled regularly for 4 hours. The antioxidant capacity of serum rose significantly and peaked at 90 minutes post wine consumption (Maxwell et al., 1994). It has also been shown in several studies that the consumption of red wine reduced the susceptibility of human plasma and low density lipoprotein to lipid peroxidation (Fuhrman et al., 1995; Duthie and Bellizzi, 1999; Whitehead et al., 1995). Purple grape juice and red wine have both been shown to have antioxidant properties(Miyagi et al., 1997). We recently demonstrated that LDL could be protected from oxidation in patients with coronary artery disease. Blood samples were obtained from the 15 patients with known stable coronary artery disease before and 14 days after they had consumed 5-7 ml/kg of purple grape juice daily. The purpose was to determine if grape juice polyphenolic compounds with their antioxidant properties, could be absorbed from the GI tract, bind to LDL, and act *in vivo* as antioxidants (Stein et al., 1999). The LDL-C was isolated from the patient's blood. The LDL-C was subjected to standard cupric ion oxidation conditions. The lag time was determined as the time required before the LDL began to be oxidized. The longer the lag time, the better the LDL is protected from oxidation. We observed that there was a significant increase in lag time from 87 minutes to 119 minutes (35%, p < 0.02) before the LDL-C began to be oxidized in the blood obtained after 14 days of grape juice consumption compared to control. This suggested that the purple grape juice had a significant antioxidant protective effect on the patient's LDL-C (Stein et al., 1999). This occurred in spite of the fact that the patients were also taking another good antioxidant, 400 IU of vitamin E daily (Stein et al., 1999). A second study of eight healthy humans, not taking vitamin E was done. They consumed 400 ml of purple grape juice for one week. The lag time increased from 150 minutes to 200 minutes after one week of consumption of purple grape juice (p < 0.05) compared to control (Vinson, 1998).

These two studies confirm that the flavonoids are absorbed, get into the blood, bind to LDL-C and can protect LDL-C from oxidation. More studies with flavonoid sources like red wine and purple grape juice need to be done for much longer periods of time to demonstrate the potential benefit from the antioxidants in the beverages.

6.2. Effects of Red Wine and Purple Grape Juice on Experimental Atherosclerosis

Many experimental animal models have been used to study the initiation and progression of atherosclerosis using rabbits, birds, dogs, rats, pigs, and nonhuman primates (Gross, 1985). A variety of interventions have been done to demonstrate a reduction in the extent of atherosclerosis produced. These include platelet inhibition, antioxidants, antineoplastic or lipid lowering agents, and a variety of substances to improve endothelial function (Bocan, 1998; van der Loo and Martin, 1997).

The complex interactions shown in Figure 2 and Figure 3 suggest that the flavonoids in grape products may inhibit the development of atherosclerosis in multiple ways. If they improve EC function and produce more NO, then platelets and monocytes are less likely to stick on the wall and release their growth factors and initiate VSMC proliferation and migration (Fig. 2). Also if more NO can be produced by dysfunctional ECs, then this also would inhibit the proliferation and migration of VSMCs. If grape flavonoids also acted as good antioxidants, then LDL might be protected from oxidation and be less likely to contribute to foam cell production. It would take many years to demonstrate in a prospective study in humans that drinking red wine or grape juice daily might slow the initiation and progression of atherosclerotic narrowing of arteries. There have been a number of retrospective epidemiological studies which have shown that the consumption of flavonoids in foods like tea, wine, fruits, and vegetables over 10-20 years have an inverse correlation with coronary artery disease (Hertog et al., 1993; Hertog et al., 1995; Keli et al., 1996; Key et al., 1996; Knekt et al., 1996). In order to determine if red wine or purple grape juice can slow the initiation and progression of atherosclerosis several studies have been done in the hypercholesterolemic rabbit model which has been utilized for many years (Eberhard, 1936). With this model one can feed a high cholesterol diet to normal rabbits, and within 3 months they will have extensive atheromas in their aorta and arterial tree. Klurfeld in 1981 studied six groups of rabbits. All were given the high cholesterol diet (Klurfeld and Kritchevsky, 1981). The six groups were given as treatment, (along with high cholesterol diet) sugar water, moderate amounts of red wine, white wine, pure ethanol, whiskey and beer in their drinking water for 3 months. The randomly tested blood alcohol level was undetectable in some rabbits and was between 0.01 and 0.02 gm/dL in the others, so the dose was indeed moderate. Only the red wine, with the highest amount of flavonoids, produced a statistically significant decrease in the amount of atherosclerosis on the intimal surface as determined by Sudan red staining (Klurfeld and Kritchevsky, 1981). HDL increased significantly in all animals consuming an alcoholic beverage, with red and white wine and pure ethanol showing the greatest increase. We recently studied 2 groups of rabbits, 10 in each group. All were given 0.5% cholesterol along with their normal rabbit chow for 48 days. At this time 10 were given 50 ml/kg per day of purple grape juice in their drinking water, and the other 10 were given the equivalent amount of sugars in purple grape juice. Both groups continued on the 0.5% cholesterol added to their food (Shanmuganayagam et al., 1999). After 96 days the animals were sacrificed. The arterial tree from heart to iliac arteries was stained with Sudan red to delineate atheroma formation and the percent of intima covered with atheroma determined. Areas with atheroma are stained with red while normal areas appear white. The stained arterial tree was scanned with a flatbed computer scanner (Umax Astra 12005, Umax Technologies Inc., Fremont, CA) to obtain a digital image. The images were processed using Adobe Photoshop and NIH imaging software to determine the ratio of stained to unstained areas. There was a 35% (p < 0.003) reduction in atheroma development in the rabbits drinking purple grape juice compared to those drinking sugar water (Shanmuganayagam et al., 1999). In another study, two groups of rabbits were fed 0.5% cholesterol along with their regular diet. In addition, one group received 1 mg/kg resveratrol dissolved in 0.05 ml/kg of 95% ethanol. The other group received just the 0.05 mL/kg of 95% ethanol as a control for 60 days. Resveratrol is a trihydroxystilbene found in grape skins and wine, andt is thought to have antiplatelet and antioxidant properties *in vitro* (Wilson et al., 1996). Thus, it was anticipated that the resveratrol treated rabbits might develop less atherosclerosis. However, the opposite occurred, i.e., the resveratrol treated rabbits had 27% (p < 0.02) more Sudan red staining than the control group. The resveratrol did not alter lipids,, liver or kidney function, or weight and general health of the rabbits. However, by some unknown mechanism the feeding of resveratrol, a wine constituent, produced more atheroma not less.

7. SUMMARY

In the dog, monkey, and human we have shown that 5 ml/kg of red wine or 5-10 ml/kg of purple grape juice but not orange or grapefruit juice inhibits platelet activity, and protects against epinephrine activation of platelets. Red wine and purple grape juice enhances platelet and endothelial production of nitric oxide (Fitzpatrick et al., 1993,Parker et al., 2000). This is thought to be one of the mechanisms whereby purple grape juice significantly improved endothelial function in 15 patients with coronary artery disease. The consumption of purple grape juice by the patients also offered increased protection against LDL cholesterol oxidation, even though all the patients were also taking another antioxidant, vitamin E, 400 IU/day. The number of people and animals in these studies was small; however, each one acted as their own control as measurements were made in each before, and then after consumption of red wine or purple grape juice. Thus these studies are thought to be significant. We feel that the results of these studies are encouraging and justify further research on larger numbers of subjects. This suggests that the flavonoids in purple grape juice and red wine may inhibit the initiation of atherosclerosis by one or more of the mechanisms described above. It will take years to fully characterize the potential benefits of daily consumption of red wine or purple grape juice for maintaining a healthy heart. Based on the existing evidence of antiplatelet and antioxidant benefits and improved endothelial function from red wine and purple grape juice, it seems reasonable to suggest that moderate amounts of red wine or purple grape juice be included among the 5-7 daily servings of fruits and vegetables per day as recommended by the American Heart Association to help reduce the risk of developing cardiovascular disease.

8. REFERENCES

Adams, A. K., Wermuth, W. O., and McBride, P. E., 1999, Antioxidant vitamins and the prevention of coronary heart disease, *Amer. Fam. Phys.* **60**:895-904.

Afanas'ev, I. B., Dorozhko, A. I., Brodskii, A. V., Kostyuk, V. A., and Potapovitch, A. I., 1989, Chelating and free radical scavenging mechanisms of inhibitory action of rutin and quercetin in lipid peroxidation, *Biochem. Pharmacol.* **38**:1763-1769.

Anderson, T. J., Uehata, A., Gerhard, M. D., Meredith, I. T., Knab, S., Delagrange, D., Lieberman, E. H., Ganz, P., Creager, M. A., and Yeung, A. C., 1995, Close relation of endothelial function in the human coronary and peripheral circulations, *JACC* **26**:1235-1241.

Aviram, M., Dankner, G., Brook, and J. G., 1990, Platelet secretory products increase low density lipoprotein oxidation, enhance its uptake by macrophages, and reduce its fluidity, *Arteriosclerosis* **10**:559-563.

Bocan, T. M., 1998, Animal models of atherosclerosis and interpretation of drug intervention studies [Review], Current *Pharmaceutical Design* **4**:37-52.

Bogaty, P., Hackett, D., Davies, G., and Maseri, A., 1994, Vasoreactivity of the culprit lesion in unstable angina, *Circulation* **90**:5-11.

Bossaller, C., Habib, G. B., Yamamoto, H., Williams, C., Wells, S., and Henry, P. D., 1987, Impaired muscarinic endothelium-dependent relaxation and cyclic guanosine-5-monophosphate formation in atherosclerotic human coronary artery and rabbit aorta, *J. Clin. Invest.* **79**:170-174.

Cardinal, D. C., and Flower, R. J., 1980, The electronic aggregometer: A novel device for assessing platelet behavior in blood, *J. Pharmacol. Methods* **3**:135-158.

Cavallini, L., Bindoli, A., and Siliprandi, N., 1978, Comparative evaluation of antiperoxidative action of silymarin and other flavonoids, *Pharmacol. Res. Commun.* **10**:133-136.

Cook, N. C., and Samman, S., 1996, Flavonoids - chemistry, metabolism, cardioprotective effects, and dietary sources, *J. Nutr. Biochem.* **7**:66-76.

Cooke, J. P., Stamler, J., Andon, N., Davies, P. F., McKinley, G., and Loscalzo, J., 1990, Flow stimulates endothelial cells to release a nitrovasodilator that is potentiated by reduced thiol, *Am. J. Physiol.* **259**:H804-12.

Das, N. P., and Ratty, A. K., 1986, Effects of flavonoids on induced non-enzymic lipid peroxidation, In *Plant Flavonoids in Biology and Medicine: Biochemical. Pharmacological and Structure-Activity Relationships,* Cody, V., Middleton, E., and Harborne, J. B., Alan R. Liss, New York, NY, pp. 243-247.

Demrow, H. S., Slane, P. R., and Folts, J. D., 1995, Administration of wine and grape juice inhibits *in vivo* platelet activity and thrombosis in stenosed canine coronary arteries, *Circulation* **91**:1182-1188.

DeWhalley, C. V., 1990, Flavonoids inhibit the oxidative modification of low density lipoproteins by macrophages, *Biochem. Pharmacol.* **39**:1743-1750.

Duthie, G. G., and Bellizzi, M. C., 1999, Effects of antioxidants on vascular health, *British Medical Bulletin* **55**:568-577.

Eberhard, T. P., 1936, Effect of alcohol on cholesterol-induced atherosclerosis in rabbits, *Arch. Pathol.* **21**:616-627.

Esterbauer, H., Gebicki, J., Puhl, H., and Jurgens, G., 1992, The role of lipid peroxidation and antioxidants in oxidative modification of LDL, *Free Radic. Biol. Med.* **13**:341-390.

Facing, R. M., Carina, M., Bombardelli, E., Morazzoni, P., and Morelli, R., 1994, Free radicals scavenging action and anti-enzyme activities of procyanidines from *Vitis vinifera, Arzneimittel-Forschung* **44**:592-601.

Fitzpatrick, D. F., Hirschfield, S. L., and Coffey, R. G., 1993, Endothelium-dependent vasorelaxing activity of wine and other grape products, *Am. J. Physiol.* **265**(2pt2):H774-H778.

Folts, J. D., 1998, Antithrombotic potential of grape juice and red wine for preventing heart attacks, *Pharm. Biol.* **36**:1-7.

Folts, J. D., Loscalzo, J., Muller, J. E., Schafer, A. I., and Willerson, J. T., 1999, A perspective on the potential problems with aspirin as an antithrombotic agent: A comparison of studies in an animal model with clinical trials, *JACC* **33**:295-303.

Folts, J. D., and Rowe, G. G., 1988, Epinephrine potentiation of *in vivo* stimuli reverses aspirin inhibition of platelet thrombus formation in stenosed canine coronary arteries, *Thromb. Res.* **50**:507-516.

Fraga, C. G., Martino, V. S., Ferraro, G. E., Coussio, J. D., and Boveris, A., 1987, Flavonoids as antioxidants evaluated by *in vitro* and *in situ* liver chemiluminescence, *Biochem. Pharmacol.* **36**:717-720.

Frankel, E. N., Kanner, J., German, J. B., Parks, E., and Kinsella, J. E., 1993, Inhibition of oxidation of human low-density lipoprotein by phenolic substances in red wine, *Lancet* **341**:454-457.

Fuhrman, B., Lavy, A., and Aviram, M., 1995, Consumption of red wine with meals reduces the susceptibility of human plasma and low-density lipoprotein to lipid peroxidation, *Am. J. Clin. Nutr.* **61**:549-554.

Gross, D. R., 1985, *Animal Models in Cardiovascular Research*, Martinus Nijhoff Publishers, Boston, MA,

Gryglewski, R. J., 1987, On the mechanism of antithrombotic action of flavonoids, *Biochem. Pharmacol.* **36**:317-322.

Hertog, M. G., Feskens, E. J., Hollman, P. C., Katan, M. B., and Kromhout, D., 1993, Dietary antioxidant flavonoids and risk of coronary heart disease: The Zutphen Elderly Study, *Lancet* **342**:1007-1011.

Hertog, M. G., Kromhout, D., Aravanis, C., Blackburn, H., Buzina, R., Fidanza, F., Giampaoli, S., Jansen, A., Menotti, A., and Nedeljkovic, S., 1995, Flavonoid intake and long-term risk of coronary heart disease and cancer in the seven countries study, *Arch. Intern. Med.* **155**:381-386.

Holvoet, P., and Collen, D., 1998, Oxidation of low density lipoprotein in the pathogenesis of atherosclerosis, *Atherosclerosis* **137**(Suppl.):S33-S38.

Itakura, H., 1999, Antiatherogenic effects of non-alcoholic ingredients in alcoholic beverages, In: *Moderate Alcohol Consumption and Cardiovascular Disease*, Congressi e Workshop NFI, Venice, Italy, October 30-31, 1999, Centro Studi dell'Alimentazione (Nutrition Foundation of Italy), (Abstr.).

Kauhanen, J., Kaplan, G. A., Goldberg, D. E., Salonen, R., and Salonen, J. T., 1999, Pattern of alcohol drinking and progression of atherosclerosis, *Arterioscler. Thromb. Vasc. Biol.* **19**:3001-3006.

Keevil, J., Osman, H. E., Reed, J., and Folts, J. D., 2000, Grape juice, but not orange or grapefruit juices, inhibit human platelet aggregation, *J. Clin. Nutr.* **130**:53-56.

Keli, S. O., Hertog, M. G., Feskens, E. J., and Kromhout, D., 1996, Dietary flavonoids, antioxidant vitamins, and incidence of stroke: The Zutphen study, *Arch. Intern. Med.* **156**:637-642.

Key, T. J., Thorogood, M., Appleby, P. N., and Burr, M. L., 1996, Dietary habits and mortality in 11,000 vegetarians and health conscious people: Results of a 17-year follow-up, *BMJ* **313**:775-779.

Kinsella, J. E., Frankel, E., German, B., and Kanner, J., 1993, Possible mechanisms for the protective role of antioxidants in wine and plant foods, *Food Technology* **47**:85-89.

Klurfeld, D. M., and Kritchevsky, D., 1981, Differential effects of alcoholic beverages on experimental atherosclerosis in rabbits, *Experimental and Molecular Pathology* **34**:62-71.

Knekt, P., Jarvinen, R., Reunanen, A., and Maatela, J., 1996, Flavonoid intake and coronary mortality in Finland - a cohort study, *BMJ* **312**:478-481.

Lenfant, C., 1999, Conquering cardiovascular disease, *JAMA* 282:2068-2070.

Loscalzo, J., 1990, A unique risk factor for atherothrombotic disease, *Atherosclerosis* **10**:672-679.

Maclure, M., 1993, Demonstration of deductive meta-analysis: ethanol intake and risk of myocardial infarction, *Epidemiologic Reviews* 15:328-351.

Maxwell, S., Cruickshank, A., and Thorpe, G., 1994, Red wine and antioxidant activity in serum, *Lancet* **344**:193-194.

Miyagi, Y., Miwa, K., and Inoue, H., 1997, Inhibition of human low-density lipoprotein oxidation by flavonoids in red wine and grape juice, *Am. J. Cardiol.* **80**:1627-1631.

Navab, M., Berliner, J. A., Watson, A. D., Hama, S. Y., Territo, M. C., Lusis, A. J., Shih, D. M., Frank, J. S., Demer, L. L., Edwards, P. A., and Fogelman, A. M., 1996, The Yin and Yang of oxidation in the development of the fatty streak. A review based on the 1994 George Lyman Duff Memorial Lecture, *Arterioscler. Thromb. Vasc. Biol.* **16**:831-842.

Neunteufl, T., Kostner, K., Katzenschlager, R., Zehetgruber, M., Maurer, G., and Weidinger, F., 1998, Additional benefit of vitamin E supplementation to simvastatin therapy on vasoreactivity of the brachial artery of hypercholesterolemic men, *J. Amer. Coll. Cardiol.* **32**:711-716.

Okumura, K., Yasue, H., Matsuyama, K., Ogawa, H., Morikami, Y., Obata, K., and Sakaino, N., 1992, Effect of acetylcholine on the highly stenotic coronary artery: difference between the constrictor response of the infarct-related coronary artery and that of the noninfarct-related artery, *JACC* **19**:752-758.

Osman, H. E., Maalej, N., Shanmuganayagam, D., and Folts, J. D., 1998, Grape juice but not orange or grapefruit juice inhibits platelet activity in dogs and monkeys, *J. Nutr.* **128**:2307-2312.

Parker, C., Deak, L., Frei, B., Folts, J. D., and Freedman, J. E., 2000, Oral consumption of purple grape juice inhibits platelet function and increases platelet-derived nitric oxide release, *JACC* **35**:267A(Abstract).

Pignol, B., Etienne, A., Crastes, de, Paulet, A., Deby, C., Mencia-Huerta, J. M., and Braquet, P., 1988, Role of flavonoids in the oxygen-free radical modulation of the immune response, in *Plant Flavonoids in Biology and Medicine II. Biochemical, Cellular and Medicinal Properties*, Cody, V., ed., Alan R. Liss, Inc., New York, NY, pp. 173-182.

Rabbani, L. E., and Loscalzo, J., 1994, The relationship between thrombosis and atherosclerosis, in *Thrombosis and Hemorrhage*, Loscalzo, J., and Schafer, A. I., eds., Blackwell Scientific Publications, Boston, MA, pp. 771-773.

Ribereau-Gayon, P., 1982, The anthocyanins of grapes and wines, in *Anthocyanins as Food Colors*, Markakis, P., ed., Academic Press, New York, NY, pp. 209-244.

Ross, R., 1986, The pathogenesis of atherosclerosis: An update, *New Engl. J. Med.* **314**:488-500.

Rubanyi, G. M., Romero, J. C., and Vanhoutte, P. M., 1986, Flow-induced release of endothelium-derived relaxing factor, *Amer. J. Physiol.* **250**:H1145-9.

Rusznyák, S., and Szent-Györgyi, A., 1936, Vitamin P: Flavonols as vitamins, *Nature* (Lond.) **138**:27

Scharf, R. E., and Harker, L. A., 1987, Thrombosis and atherosclerosis: Regulatory role of interactions among blood components and endothelium, *Blut* **55**:131-144.

Schroeder, S., Enderle, M. D., Ossen, R., Meisner, C., Baumbach, A., Pfohl, M., Herdeg, C., Oberhoff, M., Haering, H. U., and Karsch, K. R., 1999, Noninvasive determination of endothelium-mediated vasodilation as a screening test for coronary artery disease: Pilot study to assess the predictive value in comparison with angina pectoris, exercise electrocardiography, and myocardial perfusion imaging, *Amer. Heart J.* **138**:731-738.

Shanmuganayagam, D., Warner, T., and Folts, J. D., 1999, Effect of purple grape juice on platelet activity and development of atherosclerosis in hypercholesterolemic rabbits, *FASEB* **13**:239A(Abstract).

Sherry, S., 1984, Aspirin and antiplatelet drugs: The clinical approach, *Cardiovasc. Rev. Rep.* **5**:1208-1219.

Sies, H., 1993, Strategies of antioxidant defense, *Eur. J. Biochem.* **215**:213-219.

Silva, J. M., Riguld, J., Cheynier, V., Cheminat, A., and Moutounet, M., 1991, Procyanidin dimers and trimers from grape seeds. *Phytochemistry* **30**:1259-1264.

Singleton, K., 1981, Flavonoids, in *Advances in Food Research*, Childester, C. O., Mark, E. M., and Stewart, G. F., eds., Academic Press, New York, NY, pp. 149-242.

Solzbach, U., Hornig, B., Jeserich, M., and Just, H., 1997, Vitamin C improves endothelial dysfunction of epicardial coronary arteries in hypertensive patients, *Circulation* **96**:1513-1519.

Stein, J. H., Keevil, J. G., Wiebe, D. A., Aeschlimann, S., and Folts, J. D., 1999, Purple grape juice improves endothelial function and reduces the susceptibility of LDL cholesterol to oxidation in patients with coronary artery disease, *Circulation* **100**:1050-1055.

Takase, B., Uehata, A., Akima, T., Nagai, T., Nishioka, T., Hamabe, A., Satomura, K., Ohsuzu, F., and Kurita, A., 1998, Endothelium-dependent flow-mediated vasodilation in coronary and brachial arteries in suspected coronary artery disease, *Amer. J. Cardiol.* **82**:1535-9, A7-8.

Ting, H. H., Timimi, F. K., Haley, E. A., Roddy, M. A., Ganz, P., and Creager, M. A., 1997, Vitamin C improves endothelium-dependent vasodilation in forearm resistance vessels of humans with hypercholesterolemia, *Circulation* **95**:2617-2622.

van der Loo, B., and Martin, J. F., 1997, The adventitia, endothelium and atherosclerosis [Review], *Internat. J. Microcirculation: Clinical & Experimental* **17**:280-288.

van Hinsbergh, V. W., Scheffer, M., Havekes, L., and Kempen, H. J., 1986, Role of endothelial cells and their products in the modification of low-density lipoproteins, *Biochem. Biophys. Res. Commun.* **878**:49-64.

Vinson, J. A., 1998, Flavonoids in foods as *in vitro* and *in vivo* antioxidants, in *Flavonoids in The Living System*, Manthey, J. A., and Buslig, B. S., eds., Plenum Press, New York, NY, pp. 151-164.

Vinson, J. A., Jang, J., Dabbagh, Y. A., Serry, M. M., and Cai, S., 1995, Plant polyphenols exhibit lipoprotein-bound antioxidant activity using an *in vitro* model for heart disease, *J. Agric. Food Chem.* **43**:2798-2799.

Vita, J. A., Keaney, J. F., and Loscalzo, J., 1996, Endothelial dysfunction in vascular disease, in *Vascular Medicine: A Textbook of Vascular Biology and Disease*, Loscalzo, J., Creager, M. A., and Dzau, V. J., eds., Little, Brown & Co., Boston, MA, pp. 245-246.

Wagner, A. H., Kohler, T., Ruckschloss, U., Just, I., and Hecker, M., 2000, Improvement of nitric oxide-dependent vasodilation by HMG-CoA reductase inhibitors through attenuation of endothelial superoxide anion formation, *Arterioscler. Thromb. Vasc. Biol.* **20**:61-69.

Wang, H., Cao, G., and Prior, R. L., 1996, Total antioxidant capacity of fruits, *J. Agric. Food Chem.* **44**:701-705.

Waterhouse, A. L., German, J. B., Franke, E. N., Walzem, R. L., Teissedre, P. L., and Folts, J. D., 1997, The phenolic phytochemicals in wine, fruit and tea: Dietary levels, absorption and potential nutritional effects, In *Hypernutritious Foods*, Finley, J. W., Armstrong, D. J., Nagy, S., and Robinson, S. F., eds., AgScience, Inc., Auburndale, FL, pp. 219-238.

Waterhouse, A. L., and Teissedre, P., 1997, Levels of phenolics in California varietal wines, in *Wine: Nutritional and Therapeutic Benefits*, Watkins, T. R ., ed., American Chemical Society, Washington, DC, pp. 12-23.

Whitehead, T. P., Robinson, D., Allaway, S., Syms, J., and Hale, A., 1995, Effect of red wine ingestion on the antioxidant capacity of serum, *Clin. Chem.* **41**:32-35.

Wilson, T., Knight, T. J., Beitz, D. C., Lewis, D. S., and Engen, R. L., 1996, Resveratrol promotes atherosclerosis in hypercholesterolemic rabbits, *Life Sciences* **59**:PL15-21.

Yuting, C., Rongliang, Z., and Zhongjian, J., 1990, Flavonoids as superoxide scavengers and antioxidants, *Free Radic. Biol. Med.* **9**:19-21.

POLYPHENOL ANTIOXIDANTS IN CITRUS JUICES: *IN VITRO* AND *IN VIVO* STUDIES RELEVANT TO HEART DISEASE

Joe A. Vinson[1]*, Xiquan Liang[1], John Proch[1], Barbara A. Hontz[1], John Dancel[1], and Nicole Sandone[2]

1. ABSTRACT

It is well known that eating fruits and vegetables lowers the risk of chronic diseases such as heart disease and cancer. The question of what is/are the active ingredient(s) is still unresolved. The initial hypothesis was that the antioxidant vitamins were responsible. However, recently the polyphenols have been investigated since they have been found to have beneficial properties such as being strong antioxidants. We measured the polyphenol content of citrus juices by an oxidation-reduction colorimetric method (Folin) using catechin as the standard. The order was tangerine juice > grapefruit juice > orange juice. The antioxidant contribution of ascorbic acid was measured by the difference in Folin reactive content following removal by ascorbate oxidase. Ascorbate contributed 56 to 77% of the antioxidant content of orange juice, 46% of the single tangerine juice measured, and 66 to 77% of grapefruit juices. Polyphenol quality in the juices was analyzed by using the inhibition of lower density lipoprotein oxidation promoted by cupric ion, an *in vitro* model of heart disease. Quality decreased in the following order: orange juice > grapefruit juice > tangerine juice. In orange juice polyphenols accounted for 84-85% of antioxidant quality. The pure polyphenol hesperidin, which is common in juices, ascorbic acid, and the citrus juices, were not able to bind with LDL+VLDL and protect it from oxidation. In a hamster model of atherosclerosis, the juices were able to significantly inhibit atherosclerosis and lowered cholesterol and triglycerides. Ascorbic acid alone in the dose provided by the juices was found to have the same effect on atherosclerosis. However, the polyphenols in the citrus

[1] Department of Chemistry, [2]Department of Biology, University of Scranton, Scranton, PA 18510-4626. *Author to whom correspondence should be addressed. Telephone: (570) 941-7551; Fax (570) 941-7510; e-mail: vinson@UofS.edu

Flavonoids in Cell Function, edited by Béla Buslig and John Manthey.
Kluwer Academic/Plenum Publishers, New York, 2002.

juices are responsible for the hypolipemic effects. In a crossover study neither 200 mg of vitamin C alone or in orange juice had an *in vivo* antioxidant effect on either plasma or LDL+VLDL, and no hypolipemic effects in 8 normal human subjects.

2. INTRODUCTION

There is now a preponderance of evidence that consumption of fruits and vegetables provides protection against heart disease, stroke and cancer. For instance there was a highly significant negative association between consumption of total fresh fruits and vegetables and ischemic heart disease in a USA study (Rimm *et al.*, 1996). More recently strokes in males in the Framingham Heart Study were found to be decreased by eating fruits and vegetables (Gillman *et al.*, 1995). A very large study involved women in the Nurses' Health Study, in which they were examined for 14 years of follow-up. Men in the Health Professionals' Follow-up study had 8 years of follow-up. Citrus fruit and juice were major contributors to the stroke risk reduction from fruits and vegetables (Joshipura *et al.*, 1999). Drinking only 1 glass of orange juice daily may lower the risk of stroke in healthy men 25% as compared with only 11% from other fruits.

Initially it was assumed that the vitamin antioxidants, ascorbic acid, tocopherol, and the provitamin β-carotene were responsible for the protective effect. Recently other antioxidants such as flavonoids, and the broader classification polyphenols, have been investigated for possible benefit. We have recently shown that many polyphenols are stronger antioxidants than the vitamin antioxidants, using as a model the oxidation of the lower density lipoproteins LDL+VLDL, (Vinson *et al.,* 1995a). Polyphenols have also been found to enrich these lipoproteins (Vinson *et al.*, 1995b) after spiking in plasma. They thus can provide protection when the lipoproteins penetrate the endothelium of the aorta where they are subsequently oxidized (Steinberg *et al.*, 1989). *Citrus* flavonoids have recently been reviewed as to their chemistry, antioxidant activity and other biochemical properties (Benavente-García, 1997). The polyphenol composition of citrus juices is quite variable. Tangerines (*Citrus reticulata*) contain predominately flavanones such as hesperidin and narirutin (Mouly *et al.*, 1996), oranges (*Citrus sinensis*) hesperidin, and grapefruits (*Citrus paradisi*) contained naringin (Mouly *et al.*, 1994).

In addition to being antioxidants, citrus polyphenols such as naringin (Shin *et al.*, 1999) and hesperetin, and naringenin (Borradaile *et al.*, 1999) have been found to have potent *in vitro* hypolipemic activity. These compounds also have the ability to lower cholesterol in hypercholesterolemic rabbits (Kurosawa *et al.*, 2000) and rats (Lee *et al.*, 1999 and Shin *et al.*, 1999). We have found that an isolated citrus extract containing ascorbic acid was a potent hypolipemic agent and significantly inhibited atherosclerosis in a hamster model (Vinson *et al.*, 1998).

Our study was designed to determine the active ingredient(s) in citrus juices, which are the most popular fruit juices consumed in the USA. The *in vitro* antioxidant quantity and quality of three citrus juices; orange, grapefruit and tangerine, were investigated using our *in vitro* model of heart disease. These juices were then given as supplements in a hamster model of atherosclerosis. Orange juice and vitamin C were studied as a supplement to normal human subjects.

3. MATERIALS AND METHODS

3.1. Samples

Several brands of orange and grapefruit juice were obtained from local supermarkets. They were refrigerated and opened immediately before analysis. Only a single sample of tangerine juice was obtained direct from the manufacturer.

3.2. Antioxidant Quality Analysis

Phenols were measured in duplicate samples of each fruit at 750 nm using the Folin-Ciocalteau reagent diluted 5-fold before use (Sigma Chemical Co., St. Louis, MO), with catechin as the standard. Folin analysis was carried out before and after ascorbic acid was removed by adding 5 μL of L-ascorbate oxidase (Sigma, 100 Units/ml) to 50 μL of juice and incubating for 30 minutes at room temperature. Phenols plus ascorbate (total antioxidant capacity) were measured in the sample and phenols alone were measured in the sample after adding the enzyme. Ascorbate was determined by difference. It was found that ascorbate and catechin have the same molar response to this Folin reagent.

The quality of the phenol antioxidants was measured by determining the IC_{50} (the concentration to inhibit oxidation 50%) of juice samples (Vinson *et al.*, 1995a). Juices were added to the LDL+VLDL at concentrations ranging from 0.2 to 2 μM followed by a standard oxidation with cupric ions. The oxidation mixture was reacted with thiobarbituric acid and the products measured by fluorometry in butanol. A native sample without cupric ions and a blank sample without an antioxidant were also analyzed. All samples were done in duplicate.

The ability of vitamin C, pure polyphenols found in citrus juices and citrus juices themselves to enrich LDL+VLDL in plasma and protect them from subsequent oxidation was measured (Vinson *et al.*, 1995b). Tocopherol was also measured. Plasma was spiked with a methanol solution of pure polyphenols, vitamins, or diluted juices (50, 100 and 200 μM) along with a control, and equilibrated for one hour at 37^0C. The LDL+VLDL was isolated and oxidized with cupric ions under standard conditions. The kinetics of conjugated dienes formation were determined at 234 nm and the lag time (where the initial slow oxidation line converges with the rapid oxidation line) measured.

3.3. Hamster Study

Syrian Golden hamsters were obtained from Charles River Laboratories as weanlings and given rodent chow until they were ~100 g body weight and then they were divided into groups of 5-9 animals. Animals were given a diet of 0.2% cholesterol/10% coconut oil to induce hypercholesterolemia. They were divided into 11 groups with one being the control group given water containing Sweet N'Low. There were 3 juice groups given juice as the only source of water; orange, grapefruit and tangerine, given 1/10 diluted juice (low) and ½ diluted juice (high) with Sweet N'Low added for taste. In the first study a high dose of ascorbic acid was given (1700 μM). In a second study there was a control group given the high cholesterol diet and there were 2 groups of ascorbic acid; low (113 μM), medium (848 μM). These latter more closely bracketed the dose of ascorbate administered in the juice groups. The groups were monitored for weight, food consumption, and liquid consumption. After 10 weeks the animals were fasted for 18 hours and blood was taken by cardiac puncture followed by hemoperfusion. Plasma was separated from the blood and stored at -80°C until assayed. Aortas were processed and analyzed as previously described (Vinson *et al.*, 1998) and the % surface of the aorta covered with foam cells determined. Plasma cholesterol, HDL and triglycerides were colorimetrically determined using Sigma enzyme kits. Groups were statistically compared.

3.4. Single Dose Human Study

In the first study, six normal volunteers (3 females and 3 males, age 19-54) were asked to refrain from fruits and beverages containing vitamin C for 1 week before and during the entire study. Three subjects took Tang (86.6 g in 240 ml providing 200 mg of vitamin C) after an overnight fast and three who took orange juice (240 ml of Tropicana Premium Plus) which provided 200 mg of vitamin C, 2.8 mg of vitamin E, and 0.8 mg of β-carotene in addition to 97 mg of polyphenols as catechin equivalents. At 0 and 1 hour later blood samples were taken and analyzed for plasma ascorbic acid and plasma oxidizability after a 60-fold dilution of the plasma using cupric ions as the prooxidant (Regnström *et al.*, 1993). This model measures the effect of both plasma and lipid-bound antioxidants on the lipids in the lipoproteins which are oxidized. The oxidation is measured by the absorbance of conjugated dienes at 234 nm. The oxidizability is measured by the length of the lag time. The longer the lag time, the more antioxidants are present.

3.5. Short-term Human Supplement Study

Eight subjects (5 females and 3 males, aged 19-54) were given in a cross-over design with a washout period. They were given the same supplements as the single dose study except they were divided in half, with the first given during breakfast and the second during the evening meal. The supplementation lasted a week and fasting blood samples were taken before and after supplementation. A 1 week washout period followed and a fasting blood sample was then taken. The other supplement was given for 1 week and the sampling repeated. Plasma cholesterol, triglycerides, lipid peroxides and oxidizability were measured (Vinson *et al.*, 1998).

4. RESULTS

The results of the polyphenol and ascorbic acid determination along with quality measurements are found in Table 1. The average of polyphenols in the orange juices was 768±182 μM and in grapefruit juices was 1756±228 μM, more than twice the level in orange juices. Tangerine juice had the second highest concentration of polyphenols at 1380 μM. The quality of the antioxidants in orange juices was significantly better than those in grapefruits, $p < 0.005$, as measured by IC_{50}, with the averages being 0.42±0.17 and 1.05±0.15, respectively. Ascorbic acid accounted for 56-77% of the Folin antioxidants in the orange juices as measured by the Folin reaction, while in grapefruit juice it accounted for 23-34%. Tangerine juice was intermediate at 54% ascorbate. Ascorbate provided only

Table 1. Antioxidant quantity and quality in commercial citrus juices

Juice	Folin (μM)	Vitamin C (μM)	Folin w/o C (μM)	Vitamin C % of Folin	IC_{50} (μM)
orange juice from concentrate	2008	1355	643	67.5%	0.33
orange juice not from concentrate enriched in vitamins A, C & E	4077	3144	930	77.1%	0.38 0.45*
orange juice not from concentrate	1750	1117	633	63.8%	0.41
orange juice from concentrate	2268	1268	1000	55.9%	0.65
orange juice fresh squeezed	2015	1378	633	68.4%	0.34 0.41*
grapefruit juice frozen, not from concentrate	2497	580	1917	23.3%	1.16
grapefruit juice	2419	825	1594	34.1%	0.95
tangerine juice frozen, from concentrate	3000	1620	1380	54.0%	0.50

*vitamin C removed with ascorbate oxidase

15-16% of the quality of the antioxidants in orange juice as measured by IC_{50}, the rest was supplied by the polyphenols.

Enrichment experiments were done to determine if the antioxidants in the juices, either as pure compounds, or as natural mixtures in the juices, were able to bind to LDL+VLDL and protect them from *ex vivo* oxidation. The results are displayed in Fig. 1. As can be easily seen, ascorbic acid did not exhibit any appreciable lipoprotein-binding ability while hesperidin had a small activity. Tocopherol and quercetin were strong antioxidants in this model. The juices, on the contrary, were prooxidants.

In the hamster study there was no difference in weight gain or food consumption among the groups with the exception of the high ascorbic acid group which gained significantly less weight than the control, $p < 0.01$. The composition of the groups and the average fluid, ascorbate and polyphenol consumption are shown in Table 2. There was a tendency for less consumption when the juice concentration was higher. The results are listed in Table 3 for the lipid and atherosclerosis data. Cholesterol was significantly reduced with the high dose of orange juice, and medium and high doses of ascorbate. Triglycerides were significantly decreased in all the groups except the high orange juice group. Atherosclerosis was decreased in all the juice and vitamin C groups.

The single dose human study did not show any significant effect on plasma oxidizability after consuming 200 mg of vitamin C in the form of Tang or in orange juice along with polyphenols. The baseline lag time was 126±54 minutes and 1 hour later 133±59 minutes for the Tang, and 98±42 minutes baseline and 1 hour later 99±41 minutes for the orange juice. In the one week supplementation there was no significant effect on

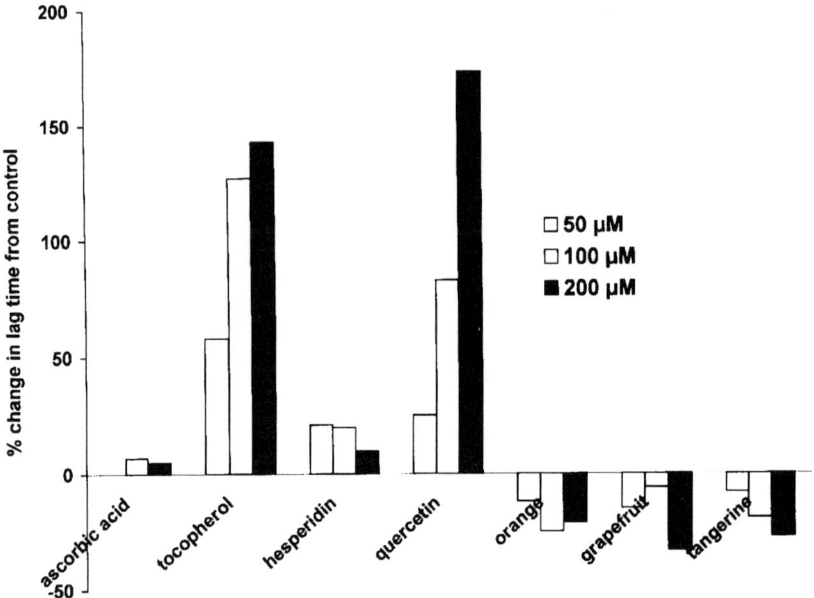

Figure 1. *Ex vivo* plasma enrichment of ascorbic acid, polyphenols and citrus and effect on oxidizability of subsequently isolated LDL+VLDL as measured by the change in lag time compared to the control

either lipids or plasma oxidizability in normal subjects as seen in Fig. 2. However, there was a significant increase in plasma ascorbate with both supplements, 60% for the Tang and 100% for the orange juice.

5. DISCUSSION

5.1. Antioxidant Quantity and Quality

The Folin reaction (an oxidation-reduction reaction) was used to determine antioxidants in the juice with or without ascorbic acid being present. The citrus juices are much lower in polyphenol antioxidants (expressed as catechin equivalents) than teas (green teas averaged 13751 µM, and black teas 21060 µM) and red wines which averaged 6192 µM (Vinson, 1998). Orange juice which averaged 768 µM was lower in polyphenols than oranges which averaged 1400 µmol/kg (Vinson, 1998).

The quality of the juice antioxidants depends on the quality of the individual antioxidants present, as well as any synergism or antagonism between them. In general, the sugar conjugated forms of the polyphenols were lower in quality than their corresponding aglycones (Vinson *et al.*, 1995a). We have also found that polyphenols are typically better antioxidants with lower IC_{50} values than the vitamins, such as vitamin C with an IC_{50} of 1.45 µM (Vinson *et al.*, 1995b). That would explain why the quality of the citrus juices

Table 2. Composition of the hamster groups, daily fluid consumption and daily dose of ascorbate and polyphenols/hamster

Group	Fluid Consumption (mL)	Ascorbate Dose (μmol)	Polyphenol Dose (μmol)
control	13.7	---	---
low ascorbic acid (113 μM)	11.7	1.3	---
medium ascorbic acid (848 μM)	8.7	7.4	---
high ascorbic acid (1700 μM)	13.5	23.0	---
low orange juice (137 μM ascorbate) (63.3 μM polyphenols)	11.9	1.6	0.8
high orange juice (683 μM ascorbate (317 μM polyphenols)	5.5	3.8	1.8
low grapefruit juice (82.5 M ascorbate) (159 μM polyphenols)	8.8	0.7	1.4
high grapefruit juice (413 μM ascorbate) (797 μM polyphenols)	4.2	1.7	3.3
low tangerine juice	11.4	1.8	1.6

were almost all lower than 1.00 μM and why, even though there is 56-77% of the moles of antioxidants in the orange juice as ascorbate, it provides only 15-16% of the quality of the antioxidants. The negative antioxidant results of the enrichment experiments suggested the probable lack of binding and/or antioxidant activity of the citrus juices' individual components such as ascorbate and hesperidin.

5.2. Hamster Study

For the juices the most comparable ascorbic acid group as determined by the ascorbate dose data in Table 2 was the low ascorbate. The sole exception being the high dose of orange juice which is midway between the low and medium ascorbate group. Thus the significant lowering of cholesterol and triglycerides found in the low dose citrus juice groups cannot be attributed to vitamin C which has no hypolipemic activity at this dose, but must be due to the polyphenols acting alone. Similar to the lack of a dose-response effect of low and medium dose ascorbate on atherosclerosis, there was no dose-response of orange and grapefruit juices on atherosclerosis. There appears to be a saturation effect at the higher dose. This was confirmed in the animal experiments with some pure citrus polyphenols mentioned in the Introduction section. The significant benefit to atherosclerosis can be attributed to the ascorbic acid in the juices.

Table 3. Average (±s.d.) hamster lipid and atherosclerosis values after 10 weeks of supplementation

Group	Cholesterol (mg/dL)	Triglycerides (mg/dL)	% of Aorta with Atherosclerosis
control	348±104	756±135	24.7±6.7 %
high ascorbic acid	175±58* -49.6%	153±56* -79.8%	12.8±5.6%* -48.1%
low ascorbic acid‡	-8.1±1.2%	+13.5±4.8%	-77.4±59.1%*
medium ascorbic acid‡	-46.6±13.7%*	-57.2±34.6%*	-81.9± 43.8%*
low orange juice	169±40* -51.1%	217±30* -71.3%	12.0±4.6%* -51.4%
high orange juice	299±141 -13.5%	565±405 -25.3%	15.3±7.0%** -38.1%
low grapefruit juice	300±66 -13.5%	566±51* -25.1%	8.9±3.2%* -64.0%
high grapefruit juice	343±67 -1.2%	351±107* -53.6%	10.4±5.1%* -57.9%
low tangerine juice	354±113 +2.0%	443±223* -41.4%	11.2±4.4%* -54.6%

‡Compared to their own control values, * p < 0.05, ** p = 0.07 vs. control.

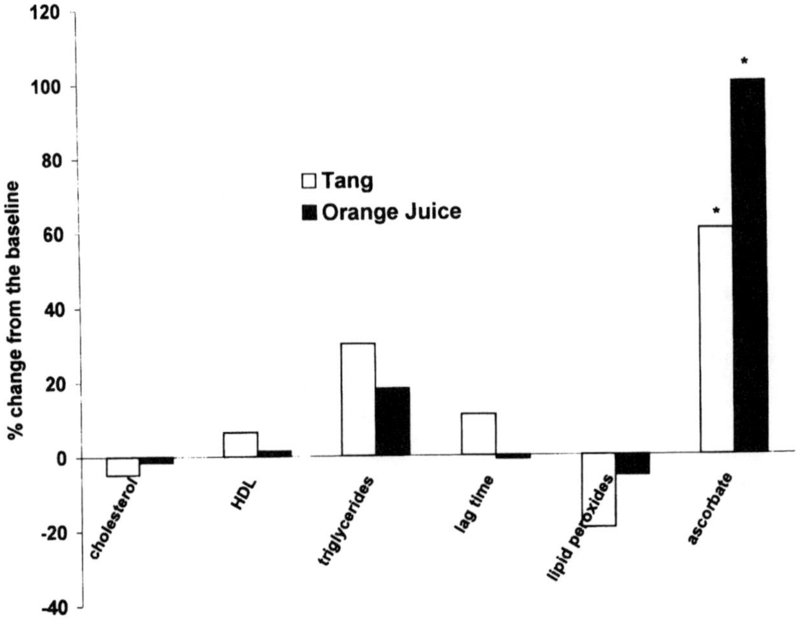

Figure 2. Effect of short-term supplementation of 200 mg of vitamin C in Tang or orange juice on normal subjects' plasma lipids, ascorbate, lipid peroxides and oxidizability. *Significantly different from baseline, p < 0.05

5.3. Human studies

There were no changes in any lipid or antioxidant parameter after either the single-dose or short-term study. In the only published study with a low dose of vitamin C, a group of smokers were given 145 mg of vitamin C as orange juice plus 16 mg of β-carotene (Abbey *et al.*, 1995). There was a 60% increase in plasma ascorbate, similar to our study. Confirming our study, Abbey found that plasma lipid peroxides were not decreased, nor did LDL lag time increase.

Citrus flavonoids such as naringenin, hesperetin and methoxy flavanone metabolites have been found in human plasma and urine after drinking grapefruit juice and orange juice (Ameer *et al.*, 1996). The lack of an *in vivo* antioxidant (Ameer *et al.*, 1996) effect appears to be due to the poor antioxidant quality of these compounds after absorption, since it is well known that citrus flavonoids are extensively conjugated *in vivo*. Another contributing factor is the lack of lipoprotein-bound antioxidant activity by citrus juices as seen in our *ex vivo* plasma enrichment study.

Interestingly even though orange juice is the major source of vitamin C in the American diet and protective for stroke (Joshipura, 1999) , consumption of C as a supplement does not have a significant preventive effect for strokes in healthy men (Ascherio *et al.*, 1999). There must be some other protective substances in citrus juices; we believe them to be polyphenols. In our human study, although there was no lipid or antioxidant effect in normal subjects, the hypolipemic effect found with hamsters suggests that citrus juices should be studied as supplements in hyperlipemic humans as a means to possibly reduce the risk of heart disease.

6. ACKNOWLEDGMENTS

Support for this work was provided by a contract from the Florida Citrus Commission.

7. REFERENCES

Abbey, M., Noakes, M., and Nestel, P. J., 1995, Dietary supplementation with orange and carrot juice in cigarette smokers lowers oxidation products in copper-oxidized low-density lipoproteins, *J. Am. Diet. Assoc.* **95**:671-675.

Ameer, B., Weintraub, R. A., Johnson, J. V., Yost, R. A., and Rouseff, R. L., 1996, Flavanone absorption after naringin, hesperidin and citrus administration, *Clin. Pharmacol. Ther.* **60**:34-40.

Ascherio, A., Rimm, E. B., Hernán M. A., Giovannucci, E., Kawachi, I., Stampfer, M. J., and Willett, W. C., 1999, Relation of consumption of vitamin E, vitamin C, and carotenoids to risk for stroke among men in the United States, *Arch. Int. Med.* **130**: 963-970.

Benavente-García, O., Castillo, J., Marin, F. R., Ortuño, A., and Del Río, J. A., 1997, Uses and properties of *citrus* flavonoids, *J. Agric. Food Chem.* **45**:4505-4515.

Borradaile, N. M., Carroll, K. K., and Kurowska, E. M. 1999, Regulation of Hepg2 cell apolipoprotein B metabolism by the citrus flavanones hesperetin and naringenin, *Lipids* **34**:591-597.

Gillman, M. W., Cupples, L. A., Gagnon, D., Posner, B. M., Ellison, R. C., Castell, and W. P., Wolff, P. A., 1995, Protective effect of fruits and vegetables on development of stroke in men, *J. Am. Med Assoc.* **273**:1113-1117.

Joshipura, K. J., Ascherio, A., Manson, J. E., Stampfer, M. J., Rimm, E. B., Speizer, F. E., Hennekens, C. H., Spiegelman, D., and Willett, W. C., 1999, Fruit and vegetable intake in relation to risk of ischemic stroke, *J. Am. Med. Assoc.* **282**:1233-1239.

Kurowska, E. M., Borradaile, N. M., and Carroll, K. K., 2000, Hypocholesterolemic effects of dietary citrus juices in rabbits, *Nutr. Res.* **20**:121-130.

Lee, S. H., Park, Y. B., Bae, K. H., Bok, S. H., Kwon, Y. K., Lee, E. S., and Choi, M. S., 1999, Hypocholesterolemic effect of naringin associated with hepatic cholesterol regulating enzyme changes in rats, *Ann. Nutr. Metab.* **43**:173-180.

Mouly, P. P., Gaydou, E. M., Arzouyan, C. R., and Estienne, J. M., 1996, Différenciation des jus de *Citrus* par analyses statistiques multivariées, Partie II. Cas des oranges et des mandarines, Analysius **24**:230-239.

Mouly, P. P., Arzouyan, C. R., Gaydou, E. M., and Estienne, J. M., 1994, Differentiation of citrus juices by factorial discriminant analysis using liquid chromatography of flavanone glycosides, *J. Agric. Food Chem.* **42**:71-79.

Regnström, J., Ström, M. P., Nilsson, J., 1993, Analysis of lipoprotein diene formation in human serum exposed to copper, *Free Rad. Res. Commun.* **19**:267-278.

Rimm, E. B., Ascherio, A., Giovanucci, E., Speigelman, Stampfer, D., and Willett, W. C., 1996:, Vegetable, fruit and cereal fiber intake and risk of coronary heart disease among men, *J. Am. Med. Assoc.* **275**:447-51.

Shin, Y. W., Bok, S. H., Jeong, T. S., Bae, K. H., Jeoung, N. H., Choi, M. S., Lee, S. H., and Park, Y. B., 1999:, Hypocholesterolemic effect of naringin associated with hepatic cholesterol regulating enzyme changes in rats, *Int. J. Vitam. Nutr. Res.* **69**:341-347.

Steinberg, D., Parathasarathy, S., Carew, T. E.; Khoo, J. C., and Witzum, J. L., 1989, Beyond cholesterol; modification of low-density lipoprotein that increases its atherogenicity, *New Engl. J. Med.* **320**:915-924.

Vinson, J. A., Dabbagh, Y. A., Serry, M. M., and Jang, J., 1995a:, Plant flavonoids, especially tea flavonols, are powerful antioxidants using an *in vitro* oxidation model for heart disease, *J. Agric. Food Chem.*, **43**:2800-2802.

Vinson, J. A., Jang, J., Dabbagh, Y. A., Serry, M. M., and Cai, S., 1995b:, Plant polyphenols exhibit lipoprotein-bound antioxidant activity using an *in vitro* model for heart disease, *J. Agric. Food Chem.* **43**:2798-2799.

Vinson, J. A., Hu, S.-J., Jung, S., and Stanski, A. M., 1998, A citrus extract plus ascorbic acid decreases lipids, lipid peroxides, lipoprotein oxidative susceptibility, and atherosclerosis in hypercholesterolemic hamsters, *J. Agric. Food Chem.* **46**:1453-1459.

Vinson, J. A., 1998, Flavonoids in foods as in vitro and in vivo antioxidants, *Adv. Exp. Med. Biol.* **439**:151-164.

INHIBITION OF COLONIC ABERRANT CRYPT FORMATION BY THE DIETARY FLAVONOIDS (+)-CATECHIN AND HESPERIDIN

Adrian A. Franke[1], Laurie J. Custer[1], Robert V. Cooney[1], Yuichiro Tanaka[1,3], Meirong Xu[2] and Roderick H. Dashwood[2]

1. INTRODUCTION

Dietary prevention is believed to be the most promising means by which incidence and recurrence of cancer can be reduced as evidenced by numerous epidemiologic studies (Hill, 1995). Flavonoids are known to inhibit cancer and heart disease as suggested by numerous cell, animal and human studies (Middleton and Kandaswami, 1994b; Das, 1990; Weisburger, 1992; Huang and Ferraro, 1992; Wattenberg, 1982; Hertog et al., 1995). Although reported to have beneficial health effects (Middleton and Kandaswami, 1994a) two subclasses of flavonoids, namely flavanones occurring abundantly in citrus fruits (Mouly et al., 1993) and (+)-catechins occurring in large amounts in pit fruits and green and black tea (Graham, 1992), emerge to be studied recently regarding their anti cancer effects (Cody et al., 1988). Most previously performed studies considered exclusively late stages of carcinogenesis such as tumor growth or invasion phenomena (Bracke et al., 1991; Kandaswami et al., 1991) and research on tea compounds considered either crude tea extracts or gallo-catechins (Mukhtar et al., 1992; Stich, 1992). Prevention of cancer by the flavonoid hesperidin or by the flavanone (+)-catechin at early stages of carcinogenesis and availability of these agents particularly on the cellular level are little investigated.

Carcinogenesis is believed to be a multistep process involving repeated DNA damage, initiation, proliferation, clonal selection and progression (Arnold et al., 1995). For selecting and ranking of potential chemopreventive chemicals inhibition of transformation has been

---

[1] Cancer Research Center of Hawaii, 1236 Lauhala Street, Honolulu, HI 96813. [2]The Linus Pauling Institute, Environmental and Molecular Toxicology, Oregon State University, Corvallis OR 97331. [3]current address: VA Medical Center, University of California at San Francisco, San Francisco, CA 94121. Corresponding author: Adrian A. Franke, Ph.D.,Phone: (808) 586-3008, Fax (808) 586-2970, e-mail: adrian@crch.hawaii.edu

Flavonoids in Cell Function, edited by Béla Buslig and John Manthey.
Kluwer Academic/Plenum Publishers, New York, 2002.

recommended due to relatively low costs, high speed, inclusion of the various stages of carcinogenesis and most importantly, due to the good prediction of *in vivo* activity. Only selected phenolic agents have been used in previous studies to determine effects at inhibition of transformation and only concentrations greater than 13 µM were applied (Leighton et al., 1992) which are levels greater than those normally reached in tissues by dietary exposure (Herrmann, 1988). Most cell models applied to test cancer preventive effects use synthetic carcinogens which are less relevant regarding cancer risk for humans (Arnold et al., 1995). In contrast, heterocyclic amines such as 2-amino-3-methylimidazo-[4,5-*f*]quinoline (IQ)[2], and related compounds are known to be very potent mutagens (Apostolides and Weisburger, 1995; Felton and Knize, 1991) and colon carcinogens (Ohgaki et al., 1991; Reddy and Rivenson, 1993; Guo et al., 1995) and play a very important role regarding human cancers due to the high content of these agents in heated and cooked meat (Turesky et al., 1994). Although animal models have been applied regarding flavonoid effects at inhibiting cancer or precancerous lesions induced by these food carcinogens (Takahashi et al., 1991) neither hesperidin nor (+)-catechin were included in these assays.

As the understanding of the processes during cancer development increases it becomes more obvious that preventive measures offer the best chances for efficient reduction of cancer incidence (Kelloff et al., 1994). For the reliable and rapid identification of promising cancer preventive agents to be used in clinical trials the application of the colonic aberrant crypt foci (ACF) assay has been strongly recommended (Wargovich et al., 1992, 1996). This assay capitalizes on the multistage process of colon carcinogenesis which is common to rats and humans. ACF contain elements of dysplasia including altered enzyme activity and oncogene mutations suggesting that the changes visualized in this assay are part of the most commonly hypothesized pathway leading to colon cancer. Also, aberrant crypts have been observed in resected human colons and therefore, these foci have been suggested to be precursors of colon cancer and are considered to be putative preneoplastic lesions and reliable intermediate biomarker for human colon cancer (Pretlow et al., 1991; Lam and Zhang, 1991; Wargovich et al., 1996). None of the potent inhibitors of neoplastic transformation identified in our earlier studies (Franke et al., 1998) including (+)-catechin and hesperidin have been tested regarding their potential effects to inhibit aberrant crypt foci formation.

2. METHODS

2.1. Performance of the Colonic Aberrant Crypt Assay

Effects of hesperidin and (+)-catechin (both from Sigma Chemical Co., St. Louis, MO) at inhibiting colonic aberrant crypt formation in the rat induced by IQ were measured as illustrated in Figs. 1 and 2 according to protocols established by us previously (Xu et al.,

---

[2] Abbreviations used: ACF, aberrant crypt foci; CAT, (+)-catechin; HD, hesperidin; HPLC, high preaaure liquid chromatography; IQ, 2-amino-3-methylimidazo[4,5-*f*]quinoline; MCA, 3-methyl cholanthrene; N-OH-IQ, 2-hydroxyamino-3-methylimidazo[4,5-*f*]quinoline

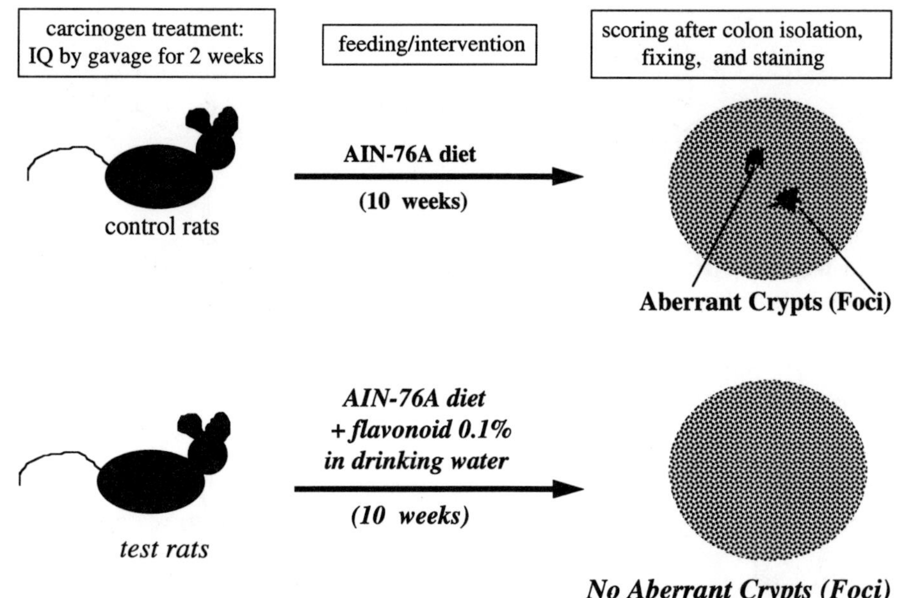

Figure 1. Performance of the colonic aberrant crypt assay. See Methods for details

1996). In brief, weaning male F344 rats were housed two to three per cage in a climate-controlled room with a 12 h light/12 h dark cycle and were given a standardized AIN-76A diet (ICN Biomedicals Inc., Aurora, OH) and water *ad libitum* during the entire experimental period. Animals were assigned randomly to the following groups (15 per group): IQ (positive control), IQ plus (+)-catechin, IQ plus hesperidin, or (5 rats/group):

All 60 male F344 rats on AIN-76A diet for entire 10 weeks

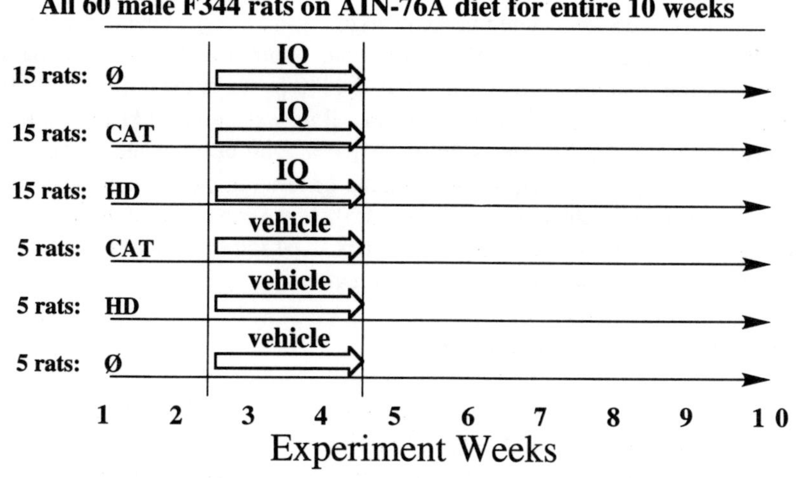

Figure 2. Treatment protocol of the colonic aberrant crypt assay. See Methods for details.

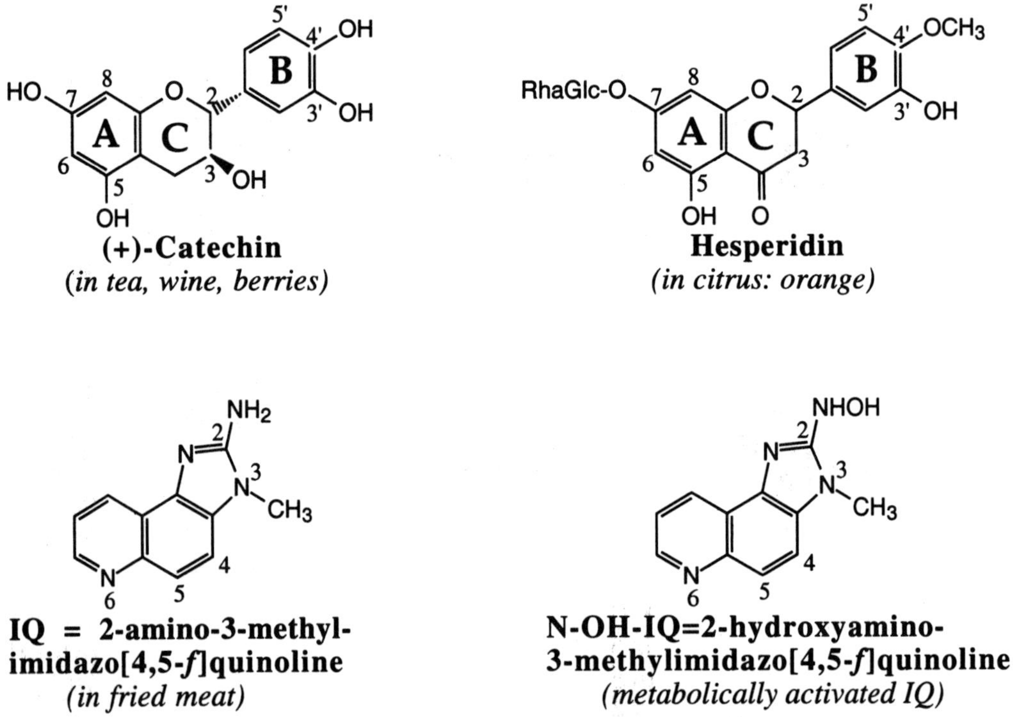

(+)-Catechin
(in tea, wine, berries)

Hesperidin
(in citrus: orange)

**IQ = 2-amino-3-methyl-
imidazo[4,5-*f*]quinoline**
(in fried meat)

**N-OH-IQ=2-hydroxyamino-
3-methylimidazo[4,5-*f*]quinoline**
(metabolically activated IQ)

Figure 3. Molecular structure and major dietary occurrence of the dietary flavonoids tested and of the dietary carcinogen IQ (2-amino-3-methylimidazo[4,5-*f*]quinoline) and N-OH-IQ (2-hydroxyamino--3-methylimidazo[4,5-*f*]-quinoline applied in the colonic aberrant crypt assay *in vivo*, or in cell transformation experiments *in vitro*

(+)-catechin alone, hesperidin alone, or vehicle alone (negative control). Food and water consumption and animal body weights were recorded throughout the 10-week experiment period. Flavonoids were administered in the drinking water at 1g/L beginning at day 1 of the assay. Periodic HPLC analysis of the drinking water revealed that the flavonoid concentration in the treatment groups was 0.09-0.11% for (+)-catechin but only 0.014-0.017% for hesperidin due to solubility problems. Rats were given 133 mg IQ/kg body weight every other day for 2 weeks starting in week 2 by oral gavage administration. Control animals were given the equivalent volume of test vehicle containing no IQ (2.5 mL/kg body weight of 55% ethanol in saline, pH 4.5). At the end of the study, animals were CO_2 asphyxiated, the colon was immediately isolated, fixed and stained followed by scoring of colonic aberrant crypts and foci as described previously.

2.2. Performance of Transformation Assay

Induction of neoplastic transformation was performed as described in detail previously (Cooney et al., 1993; Franke et al., 1998). In brief, C3H 10T1/2 murine fibroblasts were seeded at a density of 1000 cells per 60mm culture dish followed by treating the next day with either 3-methyl cholanthrene (Sigma Chemical Co., St. Louis, MO) at a final concentration of 9.3 µM for 24 hours or with IQ (Toronto Research Chemicals, North York, Ontario, Canada) at a final concentration of 10-50 µM for 24-96 hours, or with N-OH-IQ (Toronto Research Chemicals, North York, Ontario, Canada) at a final concentration of 10 µM for 24 hours, or with 10 µM IQ + male rat liver S9 (final volume 100 µL) for 24 hours according to Malaveille et al., (1989). The medium with the carcinogen was replaced after carcinogen treatment with fresh medium and the cells were grown to confluence in 8-10 days. Thereafter, cells were incubated for four to six weeks with medium containing no (positive control) or 0.1-10 µM flavonoids with weekly replenishment of the flavonoid containing culture medium. Negative control cultures were treated with vehicle alone, while positive controls were treated with the carcinogens. Four to six weeks after confluence, all cultures were fixed, stained, and scored for the presence of types II and III foci according to the guidelines described by Reznikoff et al., (1973). An average of 12 dishes per analyte and concentration were used and final results are reported as the mean of transformed foci per dish.

3. RESULTS AND DISCUSSION

3.1. Colonic Aberrant Crypt Assay Results

The mean volume of water consumption varied only slightly between the six experimental groups (23-27 mL/rat/day) however, due to the low solubility of hesperidin in the drinking water the dose per rat per day was 3.6 mg and 4.5 mg for the IQ+HD and the HD group (p for difference<0.01) while the dose for the IQ+CAT and the CAT group was 22.8 mg and 24.5 mg (p for difference=0.11), respectively (Fig. 4). Although flavonoid

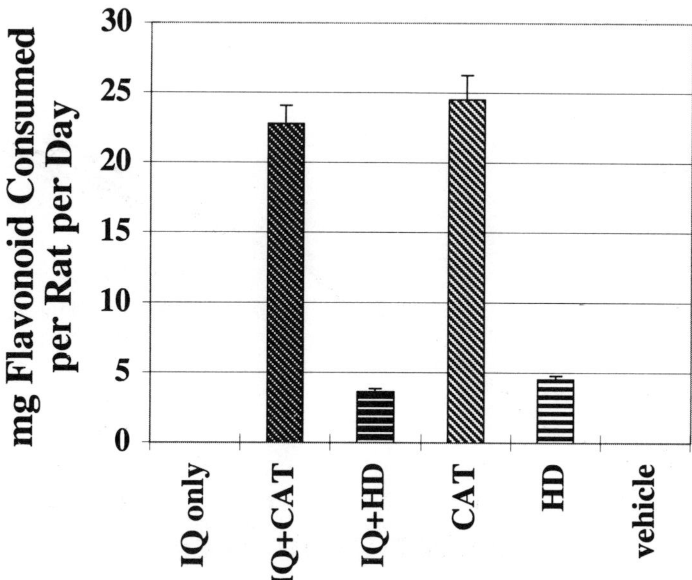

Figure 4. Mean dose of flavonoid per rat per day during the experimental 10 week period. Data were calculated from measurement of drinking water consumption and flavonoid concentration by HPLC. Error bars indicate standard error

doses were originally designed to be equal, the lower hesperidin dose was better comparable to human exposure of this flavonoid. As illustrated in Table 1, humans consume with every serving of orange juice, the food item with highest hesperidin content, approx. 1.1 mg/kg of hesperidin. Consequently, the experimental daily dose of hesperidin in the colonic aberrant crypt assay (18mg/kg) could theoretically be achieved by dietary exposure to humans. This is not true for (+)-catechin. For example, green tea, the richest food source of catechins, would lead to a catechin dose of up to 4 mg/kg per beverage serving when combining all the different catechin isomers. Daily doses of 115 mg/kg, as used in this study, could therefore not be reached in the human diet. Body weights in rats varied minimally between the 6 experimental groups (Fig. 5) and differences in mean weights of the 5 treatment groups versus the vehicle group were statistically insignificant (p>0.35).

Table 1. Food concentration and human dietary exposure of (+)-catechin and hesperidin[*]

	ppm[a]	mg/glass[b]	mg/kg[c]	Reference
(+)-catechin in green tea leaves[d]	600	3	0.05	Dalluge et al., 1998
(+)-catechin in red wine	145	35	0.5	Ghiselli et al., 1998; Burns et al., 2000
hesperidin in orange juice	320	75	1.1	Franke et al., 2000; Bronner and Beecher, 1999

[*]mean daily dose in our colonic aberrant crypt assay was 115 mg/kg for (+)-catechin and 18 mg/kg for hesperidin
[a]concentration in foods in parts per million (weight/weight)
[b]exposure per serving of an 8 oz. glass
[c]dose per serving in mg per kg body weight
[d]concentration of all catechin isomers and derivatives is 75 fold higher ((+)-catechin is 1.3% of all catechins)

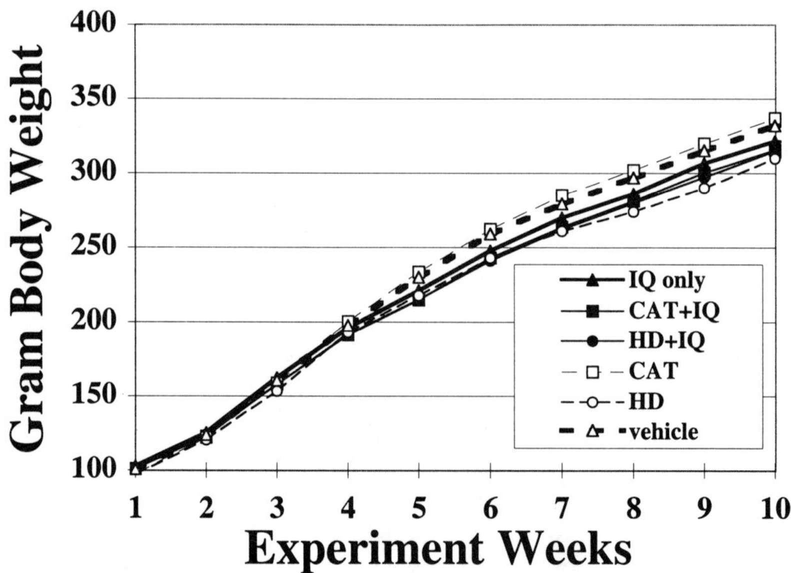

Figure 5. Body weight of rats during the 10 experimental weeks.

(+)-Catechin and hesperidin reduced the mean number of colonic aberrant crypts from 10.8 in the positive controls to 6.5 and 9.1 equivalent to a 40% and 16% inhibitory activity, respectively (Fig. 6A.) The mean number of aberrant foci was reduced by these compounds similarly (43% and 18%, respectively, Fig. 6B). Statistical significance was reached for the (+)-catechin treatment but not for the hesperidin treatment. This is probably caused by the lower hesperidin concentration applied in the drinking water (0.016% vs. 0.10% for (+)-catechin) due to the low water solubility of hesperidin (see Methods). However, when comparing the inhibitory activity after adjusting for dose the potency of hesperidin is very similar to that of (+)-catechin. The ratio of inhibitory activity per dose (measured in % inhibitory potency / % dose in drinking water) compute to 400 and 1000 for (+)-catechin and hesperidin, respectively. These values compare very favorably with the value of 635 found for the known chemopreventive agent indol-3-carbinol in tests performed by us previously applying the same aberrant crypt assay protocol (Xu et al., 1996). Consequently, we assume that higher biopotency of hesperidin could be reached by increasing the dose. The size of the foci remained unchanged as determined by the mean number of crypts per focus (Fig. 6C) which is in agreement with results on other potent dietary aberrant crypt inhibitors (Xu et al., 1996). Treatment of the animals with flavonoids alone (without IQ) resulted in no toxicity as judged by the body weights being similar to the negative control (Fig. 5) and the absence of aberrant crypt formation also observed for the negative control.

3.2. Transformation assay results

In order to avoid cost intensive animal studies it was intended to modify the 10T ½ neoplastic transformation assay as applied previously (Franke et al., 1998) to mimic the colonic aberrant crypt assay in the rat. For this reason we followed the exact transformation protocol except for replacing the carcinogen 3-methyl-cholanthrene (MCA) by IQ. Neither increasing the IQ concentration from 10 μM to 20-50 μM nor increasing the carcinogen exposure from 24 hours to 48-96 hours resulted in substantial production of neoplasticly transformed foci as observed with 10 μM MCA exposed for 24 hours. Similarly, bypassing the metabolic activation of IQ to N-OH-IQ (2-hydroxyamino-3-methylimidazo[4,5-*f*]-quinoline, see Fig. 3) by treating cells with purified N-OH-IQ at 10 μM for 24 hours was unsuccessful regarding the production of neoplastic transformation. Even the combined treatment of IQ with an microsomal liver S9 fraction according to Malaveille and coworkers (Malaveille et al., 1982, 1989) did not lead to significant production of neoplasticly transformed cell foci. Therefore, the idea had to be abandoned that the *in vivo* aberrant crypt assay can be replaced by this cell transformation assay.

4. CONCLUSION

Although major efforts were undertaken to translate the *in vivo* aberrant crypt foci system to the *in vitro* transformation system we were unsuccessful to reach neoplastic transformation using IQ or its metabolically activated carcinogenic derivatives. We showed that (+)-catechin and to a lesser extent hesperidin were able to protect against IQ-induced aberrant crypt formation. Although hesperidin showed a smaller effect at inhibiting aberrant crypt foci formation the relative effect considering dosage is equal to (+)-catechin

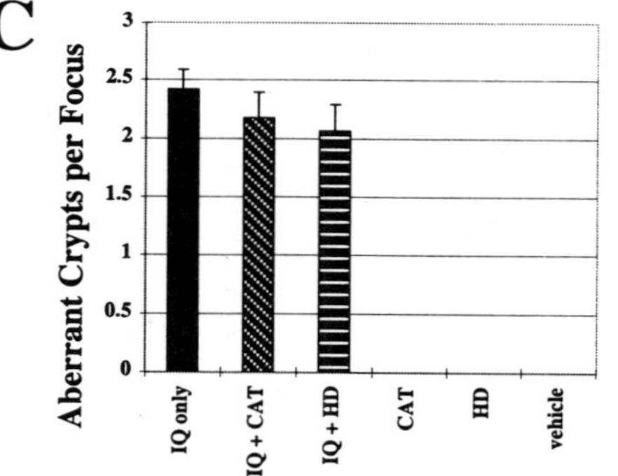

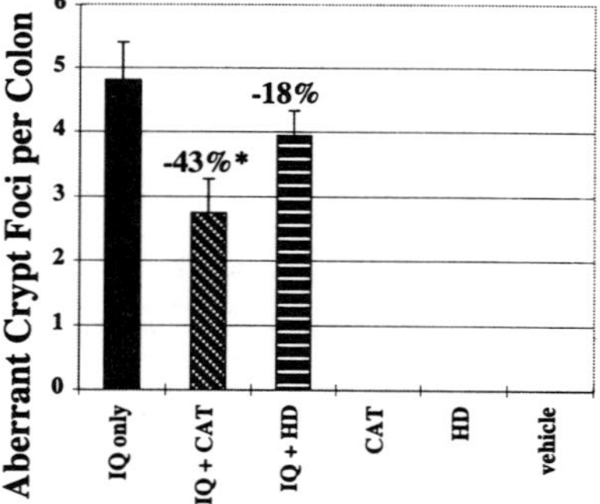

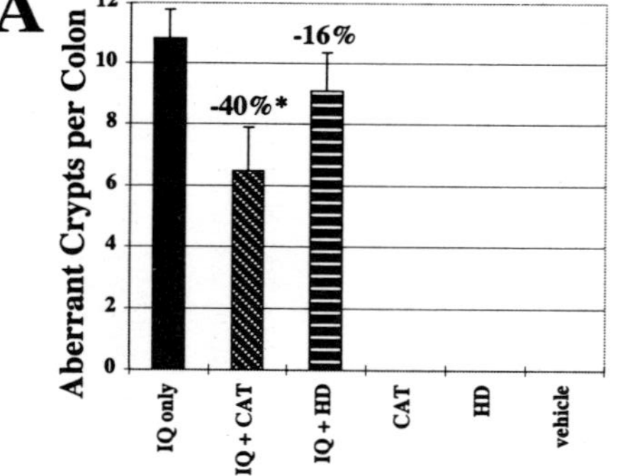

* significantly different from IQ only group (p<0.05)

Figure 6. Inhibition of IQ induced colonic aberrant crypts. A. Mean number ± standard error of aberrant crypts per rat colon; B. Mean number ± standard error of aberrant crypt foci per rat colon; C. Mean number ± standard error of ratio of aberrant crypts per focus

and other chemopreventive agents with high biopotency. Together with our earlier results showing complete inhibition of MCA induced neoplastic transformation in 10T ½ cells (Franke et al., 1998), and assuming that aberrant crypts are putative precancerous lesions, our data reported here give strong support for a protective function of the dietary flavonoids (+)-catechin and hesperidin against heterocyclic amine induced colon cancer. The potential of these flavonoids to act as useful dietary chemopreventive agents seems justified particularly in light of the good bioavailability of these agents in the circulation with catechin and hesperidin reaching plasma concentrations of 100-600 nM after green tea (Lee et al., 1995) and orange juice (Franke et al., 2000) consumption, respectively.

5. ACKNOWLEDGMENTS

This research was supported by the American Cancer Society through an Institutional Grant to the Cancer Research Center of Hawaii. Additional support was provided by grants CA65525 and CA80176. We appreciate the clerical support by Jo Ann D. Lee and the assistance of the Lab Animal Service of the University of Hawaii.

6. REFERENCES

Apostolides, Z., and Weisburger, J. H., 1995, Screening of tea clones for inhibition of PhIP mutagenicity, *Mutat. Res. Fundam. Mol. Mech. Mutagen.* **326**:219-225.

Arnold, J. T., Wilkinson, B. P., Sharma, S., and Steele, V. E., 1995, Evaluation of chemopreventive agents in different mechanistic classes using a rat tracheal epithelial cell culture transformation assay, *Cancer Research* **55**:537-543.

Bracke, M., Vyncke, B., Opdenakker, G., Foidart, J. M., De Pestel, G., and Mareel, M., 1991, Effect of catechins and citrus flavonoids on invasion *in vitro*, *Clin. Exp. Metastasis* **9**:13-25.

Bronner, W. E., and Beecher, G. R., 1999, Extraction and measurement of prominent flavonoids in orange and grapefruit juice concentrates, *J. Chromatogr. A* **705**:247-256.

Burns, J., Gardner, P. T., O'Neil, J., Crawford, S., Morecroft, I., McPhail, D. B., Lister, C., Matthews, D., MacLean, M. R., Lean, M. E. J., Duthie, G. G., and Crozier, A., 2000, Relationship among antioxidant activity, vasodilation capacity, and phenolic content of red wines, *J. Agric. Food Chem.* **48**:220-230.

Cody, V., Middleton, E., and Harborne, J. B., eds., 1988, *Plant Flavonoids in Biology and Medicine II*, Alan R. Liss, Inc., New York.

Cooney, R. V., Franke, A. A., Harwood, P. J., Hatch-Pigott, V., Custer, L. J., and Mordan, L. J., 1993, Gamma-tocopherol detoxification of nitrogen dioxide: Superiority to alpha-tocopherol, *Proc. Natl. Acad. Sci. USA* **90**:1771-1775.

Dalluge, J. J., Nelson, B. C., Thomas, J. B., and Sander, L. C., 1998, Selection of column and gradient elution system for the separation of catechins in green tea using high-performance liquid chromatography, *J. Chromatogr. A* **793**:265-274.

Das, N.P., ed., 1990, *Plant Flavonoids in Biology and Medicine III*, A. R. Liss, New York.

Felton, J. S., and Knize, M. G., 1991, Occurrence, identification, and bacterial mutagenicity of heterocyclic amines in cooked food, *Mutat. Res.* **249**:205-217.

Franke, A. A0., Custer, L. J., Cooney, R. V., Mordan, L. J., and Tanaka, Y., 1998, Inhibition of neoplastic transformation and bioavailability of dietary phenolic agents, In *Flavonoids in the Living System*, Manthey, J. A., and Buslig, B. S. eds., Plenum Press, New York, pp. 237-248.

Franke, A. A., Cooney, R. V., Henning, S. M., Okinaka, L. A., and Custer, L. J., 2000, Bioavailability and antioxidant effects of orange juice components in humans, *FASEB Journal* **14**:A491.

Ghiselli, A., Nardini, M., Baldi, A., and Scaccini, C., 1998, Antioxidant activity of different phenolic fractions separated from an Italian red wine, *J. Agric. Food Chem.* **46**:361-367.

Graham, H. N., 1992, Green tea composition, consumption, and polyphenol chemistry, *Prev. Med.*, **21**:334-350.

Guo, D., Horio, D. T., Grove, J., and Dashwood, R. H., 1995, 2-amino-3-methylimidazo[4,5-*f*]quinoline (IQ)-induced tumorigenesis in rats given chlorophyllin in the drinking water, *Cancer Lett.* **95**:161-165.

Herrmann, K., 1988, On the occurrence of flavonol and flavone glycosides in vegetables, *Z. Lebensm. Unters. Forsch.* **186**:1-5.

Hertog, M. G., Kromhout, D., Aravanis, C., Blackburn, H., Buzina, R., Fidanza, F., Giampaoli, S., Jansen, A., Menotti, A., and Nedeljkovic, S., 1995, Flavonoid intake and long-term risk of coronary heart disease and cancer in the Seven Countries Study, *Arch. Intern. Med.* **155**:381-386.

Hill, M. J., 1995, Diet and Cancer: a review of scientific evidence, *Eur. J. Cancer Prev.* **4**(Suppl. 2):3-42.

Huang, M. -T., and Ferraro, T., 1992, Phenolic compounds and cancer prevention, In *Phenolic Compounds and Their Effects on Health II*, Huang, M.-T., Ho, C.-T., and Lee, C. Y., eds., American Chemical Society, Washington, DC, pp. 8-34.

Kandaswami, C., Perkins, E., Soloniuk, D. S., Drzewiecki, G., and Middleton, E., Jr., 1991, Antiproliferative effects of citrus flavonoids on a human squamous cell carcinoma *in vitro*, *Cancer Letters* **56**:147-152.

Kelloff, G. J., Boone, C. W., Crowell, J. A., Steele, V. E., Lubet, R., and Sigman, C. C., 1994, Chemopreventive drug development: Perspectives and progress, *Cancer Epidemiol. Biomarkers Prev.* **3**:85-98.

Lam, L. K. T., and Zhang, J., 1991, Reduction of aberrant crypt formation in the colon of CF1 mice by potential chemopreventive agents, *Carcinogenesis* **12**:2311-2315.

Lee, M.-J., Wang, Z.-Y., Li, H., Chen, L., Sun, Y., Gobbo, S., Balentine, D. A., and Yang, C. S., 1995, Analysis of plasma and urinary tea polyphenols in human subjects, *Cancer Epidemiol. Biomarkers Prev.* **4**:393-399.

Leighton, T., Ginther, C., Fluss, L., Harter, W. K., Cansado, J., and Notario, V., 1992, Molecular characterization of quercetin and quercetin glycosides in *Allium* vegetables, In *Phenolic Compounds in Foods and Their Effects on Health II*, Huang, M. -T., Ho, C. -T., and Lee, C. Y., eds., American Chemical Society, Washington, DC, pp. 220-238.

Malaveille, C., Brun, G., Kolar, G., and Bartsch, H., 1982, Mutagenic and alkylating activities of 3-methyl-1-phenyltriazenes and their possible role as carcinogenic metabolites of the parent dimethyl compounds, *Cancer Res.* **42**:1446-1453.

Malaveille, C., Vineis, P., Estéve, J., Ohshima, H., Brun, G., Hautefeuille, A., Gallet, P., Ronco, G., Terracini, B., and Bartsch, H., 1989, Levels of mutagens in the urine of smokers of black and blond tobacco correlate with their risk of bladder cancer, Carcinogenesis 10:577-586.

Middleton, E., Jr., and Kandaswami, C., 1994a, Potential health-promoting properties of citrus flavonoids, *Food Technology* **48**:115.

Middleton, E., Jr., and Kandaswami, C., 1994b, The impact of plant flavonoids on mammalian biology: Implications for immunity, inflammation and cancer, In *The Flavonoids: Advances in Research Since 1986*; Harborne, J. B., ed.,.Chapman & Hall, London, pp. 619-652.

Mouly, P., Gaydou, E. M., and Estienne, J., 1993, Column liquid chromatographic determination of flavanone glycosides in citrus, Application to grapefruit and sour orange juice adulteration, *J. Chromatogr. A* **634**:129-134.

Mukhtar, H., Wang, Z. Y., Katiyar, S. K., and Agarwal, R., 1992, Tea components: antimutagenic and anticarcinogenic effects, *Preventive Medicine 21*:351-360.

Ohgaki, H., Takayama, S., and Sugimura, T., 1991, Carcinogenicities of heterocyclic amines in cooked food, *Mutat. Res.* **259**:399-410.

Pretlow, T. P., Barrow, B. J., Aston, W. S., O'Riordan, M. A., Jurcisek, J. A., and Stellato, T. A., 1991, Aberrant crypts: Putative preneoplastic foci in human colonic mucosa, *Cancer Res.* **51**:1564-1567.

Reddy, B., and Rivenson, A., 1993, Inhibitory effect of *Bifidobacterium logum* on colon, mammary, and liver carcinogenesis induced by 2-amino-3-methylimidazo[4,5-*f*]quinoline, a food mutagen, *Cancer Res.* **53**:3914-3918.

Reznikoff, C. A., Bertram, J. S., Brankow, D. W., and Heidelberger, C., 1973, Quantitative and qualitative studies on chemical transformation of clones C3H mouse embryo cells sensitive to postconfluence inhibiton of cell division, *Cancer Res.* **33**:3239-3249.

Stich, H. F., 1992, Teas and tea components as inhibitors of carcinogen formation in model systems and man, *Preventive Medicine* **21**:377-384.

Takahashi, S., Ogawa, K., Ohshima, H., Esumi, H., Ito, N., and Sugimura, T., 1991, Induction of aberrant crypt foci in the large intestine of F344 rats by oral administration of 2-amino-1-methyl-6-phenylimidazo[4,5- b]pyridine, *Jpn. J. Cancer Res.* **82**:135-137.

Turesky, R. J., Gross, G. A., Stillwell, W. G., Skipper, P. L., and Tannenbaum, S. R., 1994, Species differences in metabolism of heterocyclic aromatic amines, human exposure, and biomonitoring, *Environ. Health Perspect.* **102**(Suppl. 6):47-51.

Wargovich, M. J., Harris, C., Chen, C. -D., Palmer, C., Steele, V. E., and Kelloff, G. J., 1992, Growth kinetics of chemoprevention of aberrant crypts in the rat colon, *J. Cell Biochem.* **16G**:51-54.

Wargovich, M. J., Chen, C. -D., Jiminez, A., Stele, V., Velasco, M., Stephens, C., Price, R., Gray, K., and Kelloff, G. J., 1996, Aberrant crypt as a biomarker for colon cancer: evaluation of potential chemopreventive agents in the rat, *Cancer Epidemiol. Biomarkers Prev.* **5**:355-360.

Wattenberg, L. W., 1982, Inhibition of chemical carcinogens by minor dietary components, In *Molecular Interrelations of Nutrition and Cancer*; Arnott, M. S., VanEys, J., and Wang, Y. M., eds., Raven Press, New York, pp. 43-56.
Weisburger, J. H., 1992, Mutagenic, carcinogenic, and chemopreventive effects of phenols and catechols: The underlying mechanisms, In *Phenolic Compounds in Food and Their Effects on Health II*, Huang, M. -T., Ho, C. -T. and Lee, C. Y., eds., American Chemical Society Press, Washington, DC, pp. 35-47.
Xu, M., Bailey, A. C., Henaez, J. F., Taoka, C. R., Schut, H. A. J., and Dashwood, R. H., 1996, Protection by green tea, black tea, and indole-3-carbinol against 2-amino-3-methylimidazo[4,5-*f*]quinoline-induced DNA adducts and colonic aberrant crypts in the F344 rat, *Carcinogenesis* 17:1429-1434.

THE CITRUS METHOXYFLAVONE TANGERETIN AFFECTS HUMAN CELL-CELL INTERACTIONS

Marc E. Bracke[1]*, Tom Boterberg[1], Herman T. Depypere[2],
Christophe Stove[1], Georges Leclercq[3] and Marc M. Mareel[1]

1. ABSTRACT

Two effects of the citrus methoxyflavone tangeretin on cell-cell interactions are biologically relevant. Firstly, tangeretin upregulates the function of the E-cadherin/catenin complex in human MCF-7/6 breast carcinoma cells. This leads to firm cell-cell adhesion and inhibition of invasion *in vitro*. Secondly, tangeretin downregulates the interleukin-2 receptor on T-lymphocytes and natural killer cells. This leads to a decrease in the cytotoxic competence of these immunocytes against cancer cells. The second effect can become important when high doses of tangeretin are combined with adjuvant tamoxifen treatment for breast cancer. Experiments with nude mice bearing MCF-7/6 tumors showed that tangeretin given orally at high doses, abrogated the therapeutic suppression of tumor growth exerted by tamoxifen. No evidence for a tumor promoting effect of tangeretin by itself was found in these experiments. Tangeretin may be an interesting molecule for application in cases where immunosuppression could be clinically beneficial.

2. INTRODUCTION

Growth and invasion are the main characteristics of malignant tumors. Both activities are progressive in time and space, and are lethal for the untreated host. Growth is the net result of the balance between proliferation and elimination of tumor cells. Nowadays, oncologists have at their disposal efficient tools to control growth: radiotherapy and

[1] Laboratory of Experimental Cancerology, Department of Radiotherapy, Nuclear Medicine and Experimental Cancerology, [2]Department of Gynecology and Obstetrics, and [3]Laboratory of Bacteriology and Virology, Ghent University Hospital, De Pintelaan 185, B-9000 Gent, Belgium. *Corresponding author: Marc E. Bracke, Tel. 32 9 240 3073, Fax. 32 9 240 4991; e-mail: brackemarc@hotmail.com

Flavonoids in Cell Function, edited by Béla Buslig and John Manthey.
Kluwer Academic/Plenum Publishers, New York, 2002.

chemotherapy are meant to inhibit cell proliferation, while immunotherapy is aimed at the selective destruction of tumor cells by the patient's activated immune defense system. Invasion, however, has become a major challenge in cancer research, since not only are the mechanisms incompletely understood, but anti-invasive drugs are largely lacking (Mareel *et al.*, 1996b). A number of target features of invasive tumor cells is now emerging for drug modulation, and one of these is the loss of stable cell-cell contacts. In normal epithelial tissues, from which about 80% of the human malignant tumors (the so-called carcinomas) evolve, tissue integrity is maintained by the E-cadherin/catenin complex, which links the cells to one another (Kemler, 1992). Downregulation of the expression or the function of this complex is a common denominator of carcinomas: this phenomenon enables tumor cells to disrupt their adhesion to neighboring cells, leave their normal tissue boundaries and metastasize to distant organs. So, the E-cadherin/catenin complex is considered as an invasion suppressor, and agents that can upregulate its function are potentially useful as anti-invasive treatments (Mareel *et al.*, 1996a; Vleminckx *et al.*, 1991).

3. TANGERETIN CAN UPREGULATE THE FUNCTION OF THE E-CADHERIN /CATENIN COMPLEX

Tangeretin is a methoxyflavone accumulating in the peel oil of citrus fruits. It was tested in a screening program for the detection of anti-invasive compounds among flavonoids and polyphenolics. The assay was based on the invasion of human breast carcinoma cells (MCF-7/6) into a fragment of normal tissue *in vitro*. MCF-7/6 cells possess a complete, but functionally inactive E-cadherin/catenin complex, which is, however, amenable to upregulation. In contrast with several hundreds of compounds tested, tangeretin inhibited the invasion of MCF-7/6 cells, when added to the culture medium at a concentration of 1×10^{-5} M. This activity coincided with an enhanced aggregation by the cells, and using neutralizing antibodies, we were able to show that this aggregation was mediated by a functionally upregulated E-cadherin/catenin complex (Bracke *et al.*, 1994a). In addition to its anti-invasive effect, tangeretin possessed an anti-proliferative activity on the tumor cells (Fig. 1).

The observations with tangeretin are similar to those with tamoxifen, a synthetic antiestrogen, that is now prescribed worldwide as a successful adjuvant therapy and even as a preventive agent in human breast cancer. Since tamoxifen at 10^{-6} M also upregulated the function of the E-cadherin/catenin complex, and inhibited invasion and growth of MCF-7/6 cells *in vitro* (Bracke *et al.*, 1994b), we decided to combine tamoxifen and tangeretin. An additive effect of both compounds was found, when present in an aggregation assay with MCF-7/6 cells. These data *in vitro* were the rationale for further experiments *in vivo* in which tumor-bearing animals were expected to benefit from a combined treatment with tamoxifen and tangeretin, compared to tamoxifen alone.

Nude mice were inoculated with MCF-7/6 cells in the mammary fat pad, and divided into 4 treatment groups that received tamoxifen (3×10^{-5} M) or tangeretin (1×10^{-4} M) or tamoxifen *plus* tangeretin or only the solvent in their drinking water. When observed over a period of 16 weeks, the solvent- and tangeretin-treated mice developed tumors of similar size, while tumor growth was completely suppressed in the tamoxifen-treated group. Unexpectedly, in the combined treatment group the tumor suppressing effect of tamoxifen

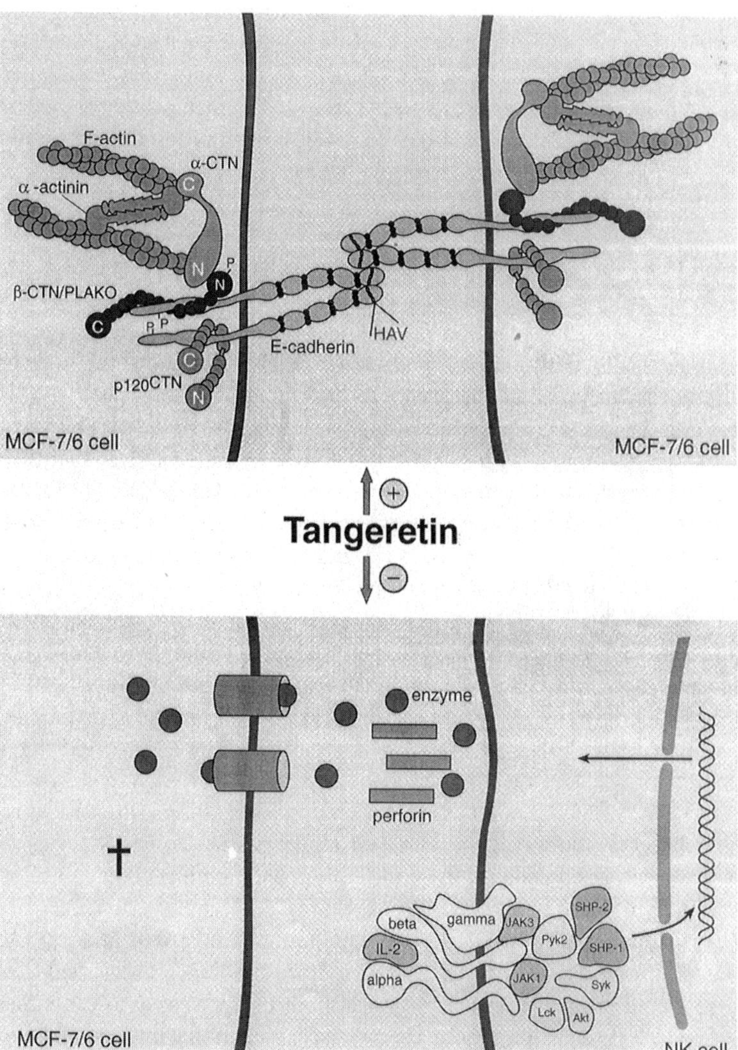

Figure 1. Effects of tangeretin on E-cadherin-mediated adhesion between human MCF-7/6 breast carcinoma cells (upper panel) and on the interaction between MCF-7/6 cells and activated natural killer (NK) cells (lower panel). In the upper panel the E-cadherin/catenin complex is shown, consisting of the transmembrane E-cadherin molecule forming parallel dimers on the surface of the epithelial cell, and interacting with a dimer from an adjacent epithelial cell in an antiparallel way. For this homophilic interaction the histidine-alanine-valine (HAV) amino acid sequence acts as the recognition site. The different catenins (CTN and PLAKO) and α-actinin link E-cadherin to the actin cytoskeleton (F-actin). N and C represent the amino and carboxy terminals of the catenins respectively; P indicates tyrosine phosphorylation sites. The lower panel shows the interleukin-2 (IL-2) receptor consisting of three transmembrane chains (α, β and γ) and signaling towards the nucleus *via* a number of tyrosine kinases and phosphatases (JAK: Janus kinase; Pyk: proline-rich tyrosine kinase; Lck: lymphocyte specific protein tyrosine kinase; SHP: Src homology 2 domain-containing phosphotyrosine phosphatase; Syk: spleen tyrosine kinase; Akt: a serine/threonine kinase). After stimulation with IL-2 the NK-cell becomes a lymphokine-activated killer (LAK) cell, and is competent to release perforin and enzymes during confrontation with a tumor cell. Due to perforin polymerization, channels are formed in the cell membrane of the target cell, and these allow the influx of cytotoxic enzymes into its cytoplasm. Tangeretin stimulates E-cadherin-mediated cell-cell adhesion, and inhibits natural killer activity.

was not augmented but abrogated by the addition of tangeretin (Bracke *et al.*, 1999). Moreover, the effect of tamoxifen alone was reversible: after 36 weeks of suppression, tumor growth resumed upon addition of tangeretin to the tamoxifen-containing drinking water. Tangeretin addition also affected the survival of the mice: while the median survival time after tumor cell inoculation was 56 weeks for the tamoxifen-treated group, this was reduced to 14 weeks in the combined treatment group.

4. TANGERETIN CAN DOWNREGULATE NATURAL KILLER CELL ACTIVITY

The action mechanism of the surprising effect of tangeretin on tumor growth *in vivo* is presumably due to an inhibition of the immune defense system of the mice against the MCF-7/6 tumors. In the blood of mice treated with tangeretin a lower number of leukocytes was counted than in controls, and this could be attributed to a selective downregulation of the lymphocyte concentration. Since tumor rejection in nude mice is known to depend mainly on one class of lymphocytes, namely the natural killer (NK) cells, and to a minor extent to T-cytotoxic lymphocytes, we further studied the effect of tangeretin on these killer cells *in vitro*. The killer activity was assessed in a ^{51}chromium release assay, in which the killer cells are confronted with ^{51}chromium-loaded target cells, and their activity is proportional to the amount of radioactive label released from the target cells into the culture medium. NK and T-cytotoxic lymphocytes were activated with interleukin-2 (IL-2) before the start of the assay, in order to obtain "lymphokine activated killer" (LAK) cells, and treated or not with tangeretin (1×10^{-4} M for 24 h). Both with murine and human LAK cells tangeretin reduced the killing activity to about half of the untreated-control value. Furthermore, ongoing experiments have shown that some other polymethoxylated flavones (e.g. nobiletin) are even more potent than tangeretin in inhibiting killer cell activity.

We speculate that the molecular target of tangeretin in the lymphocytes is the IL-2 receptor. This receptor is crucial for the proliferation and immune competence of NK cells and T-cytotoxic lymphocytes upon triggering with IL-2. This trimeric receptor is built up by an α chain (binding to the IL-2 ligand with low affinity), a β and a γ chain (both responsible for the establishment of a high-affinity binding). The β and γ chains are implicated in signal transduction via cytoplasmic tyrosine kinases towards transcription factors (Roitt *et al.*, 1998). Flow cytometry data have shown that tangeretin downregulates the expression of the β and γ chains, but not of the α chain. So, tangeretin may interfere with normal immune defense mechanisms by making killer cells insensitive to ambiant IL-2, their natural trigger for activation (Fig.1).

5. CONCLUSION

In conclusion, tangeretin has multiple effects on cancer. It possesses the intrinsic potential to activate the E-cadherin/catenin complex in human MCF-7/6 breast carcinoma cells, rendering them non-invasive. *In vivo,* however, this beneficial effect is overshadowed by the suppression of the host's immune defense against the tumor, and this phenomenon neutralizes the growth inhibitory effect of tamoxifen. We want to state explicitly that we have no data to support the idea that drinking citrus juices or eating citrus fruits can provoke any harm to breast cancer patients under tamoxifen: the doses of tangeretin ingested in this way are one hundred times lower then the doses in our animal experiments, and citrus products certainly contain many health promoting compounds. We would only warn these patients for dietary supplements of citrus methoxyflavones, that sometimes contain 1 gram of tangeretin. Those pills may interfere with tamoxifen in an undesired way. Due to its downregulation of the IL-2 receptor on immune cells, on the other hand, tangeretin may develop as an useful immunosuppressive agent in transplantation medicine, in autoimmune diseases or in the treatment of lymphomas that express the receptor excessively. So, interest may grow from its unique mechanism of downregulation of the IL-2 receptor, which is different from the mechanism of cyclosporin A, one of the reference drugs in transplantation biology.

6. ACKNOWLEDGMENTS

Part of this work was supported by the State of Florida Department of Citrus, and by the Sportvereniging tegen Kanker, Brussels, Belgium. Tom Boterberg and Christophe Stove are research assistants of the FWO (Fund for Scientific Research, Flanders, Belgium).

7. REFERENCES

Bracke, M. E., Bruyneel, E. A., Vermeulen, S. J., Vennekens, K., Van Marck, V., and Mareel, M. M., 1994a, Citrus flavonoid effect on tumor invasion and metastasis, *Food Technology* **48**:121-124.

Bracke, M. E., Charlier, C., Bruyneel, E. A., Labit, C., Mareel, M. M., and Castronovo, V., 1994b, Tamoxifen restores the E-cadherin function in human breast cancer MCF-7/6 cells and suppresses their invasive phenotype, *Cancer Res.* **54**:4607-4609.

Bracke, M. E., Depypere, H. T., Boterberg, T., Van Marck, V. L., Vennekens, K. M., Vanluchene, E., Nuytinck, M., Serreyn, R. and Mareel, M. M., 1999, Influence of tangeretin on tamoxifen's therapeutic benefit in mammary cancer, *J. Natl. Cancer Inst.* **91**:354-359.

Kemler, R., 1992, Classical cadherins, *Semin. Cell Biol.* **3**:149-155.

Mareel, M., Berx, G., Van Roy, F., and Bracke, M., 1996a, Cadherin/catenin complex: a target for antiinvasive therapy? *J. Cell. Biochem.* **61**:524-530.

Mareel, M., Noë, V., Vermeulen, S., and Bracke, M., 1996b, Anti-invasive therapy: manipulation of the E-cadherin/catenin/cytoskeleton complex, *Anticancer Drugs* **7**(Suppl 3):149-156.

Roitt, I., Brostoff, J., and Male, D., 1998, Cell-mediated immune reactions, In: *Immunology,* 5th Ed., Roitt, L, Brostoff, J., and Male, D., Eds., Mosby International Ltd, London, pp. 122-123.

Vleminckx, K., Vakaet, L. Jr., Mareel, M., Fiers, W., and Van Roy, F., 1991, Genetic manipulation of E-cadherin expression by epithelial tumor cells reveals an invasion suppressor role, *Cell* **66**:107-119.

XANTHINE OXIDASE AND XANTHINE DEHYDROGENASE INHIBITION BY THE PROCYANIDIN-RICH FRENCH MARITIME PINE BARK EXTRACT, PYCNOGENOL®: A PROTEIN BINDING EFFECT

Hadi Moini[1,2], Qiong Guo[1,3], and Lester Packer[1,2]

1. INTRODUCTION

Polyphenolic compounds have recently attracted attention due to their ability to scavenge reactive oxygen species. Therefore, flavonoids alone or as a part of a complex mixture of polyphenols have been used therapeutically in European countries and in the US as dietary supplement (Kottke, 1998). Pycnogenol® , an extract from French maritime pine bark (PBE), is an example of a highly standardized mixture of certain polyphenolic compounds. Procyanidins formed by catechin and epicatechin units with a degree of polymerization of up to heptamer are major constituents of PBE and constitute 75% of its weight. PBE also contains monomer flavonoids such as catechin and taxifolin and phenolcarbonic acids and their glycosides. Recently, various biochemical and pharmacological studies have revealed some interesting properties of PBE, both *in vitro* and *in vivo* (Packer et al., 1999).

One of the underlying mechanisms for the observed biological effects of PBE might be the interaction with cellular proteins namely, modulation of enzyme activities. However, the biochemical basis of PBE action on enzyme activities is not known. Due to the antioxidant activity of its constituents, PBE might change the redox state of enzymes, particularly those, which are redox sensitive, or react with free radicals generated at the active site

[1] Department of Molecular and Cell Biology, 251 Life Sciences Addition, University of California at Berkeley, Berkeley, CA 94720-3200, USA; [2]Present address: Department of Molecular Pharmacology & Toxicology, School of Pharmacy, University of Southern California, 1985 Zonal Ave., Los Angeles, CA 90033, Phone: (323) 442-2770, FAX: (323) 224-7473; [3]National Institute of Environmental Health Sciences, P. O. Box 12233, MD FO-02, 111 T. W. Alexander Dr.ive, Research Triangle Park, NC 27709.

Flavonoids in Cell Function, edited by Béla Buslig and John Manthey.
Kluwer Academic/Plenum Publishers, New York, 2002.

of the enzyme, thereby affecting enzyme activities. It was proposed that the redox activity of monomeric flavonoids is responsible for their effects on enzyme activities (Elliott et al., 1992; Kemal et al., 1987). It is well known that procyanidins bind with high affinity to extended proteins rich in proline and to histatins (Hagerman and Butler, 1981; Olson et al., 1974). Therefore, an alternate mechanism to modulate enzyme activities might be binding of procyanidin constituents of PBE to the enzyme, changing its conformation and thus affecting its activity. In most studies, the inhibitory action of procyanidins on enzyme activity has been attributed to their ability to bind to proteins. Such binding has been indirectly shown by the lack of correlation between enzyme inhibition and the binding of related polyphenols to hemoglobin (Hataro et al., 1990).

Therefore, possible involvement of the redox activity and/or direct protein binding property of PBE in its action on the activities of some purified enzymes has been evaluated.

2. METHODS[2]

2.1. Measurement of Enzyme Activities

Enzyme activity measurements were performed in 0.1 M phosphate buffer, pH 7.4. Xanthine oxidase (XO) and xanthine dehydrogenase (XDH) activities were determined spectrophotometrically by following uric acid production at 295 nm and β-NAD$^+$ reduction at 340 nm, respectively. The reactions were started by addition of either 12 mU/mL XO or 6 mU/mL XDH. Glucose oxidase (GO) and ascorbate oxidase (AO) activities were measured polarographically. The reactions were started by addition of either 110 mU/mL glucose oxidase or 30 mU/mL ascorbate oxidase.

2.2. Polyacrylamide Gel Electrophoresis (PAGE)

Discontinuous PAGE was carried out under either denaturing or non-denaturing conditions as described by Laemmli for denaturing condition (Laemmli, 1970). In non-denaturing gel electrophoresis, sodium dodecyl sulfate (SDS), reducing agent (2-mercapto-ethanol), and stacking gel were omitted. GO (1.3 µg protein) were subjected to PAGE under either non-denaturing or denaturing conditions on a 10% gel in the presence and absence of PBE. XO (9.25-12.6 µg protein) was subjected to PAGE under non-denaturing condition on a 3-8% NuPAGE® precast gel, and denaturing condition on a 10% gel in the presence and absence of test compounds. Following electrophoresis, the gels were stained either for activity or protein. Activity staining of XO was performed in 50 mM Tris/HCl, pH 7.6, containing 0.5 mM xanthine, and 0.25 mM nitroblue tetrazolium (NBT) (Ozer et al., 1998). Staining of gels was continued until the activity band(s) appeared on the gels (4-7 min).

---

[2] Abbreviations AO: ascorbate oxidase; BSA: bovine serum albumin; DMPO: 5, 5-dimethyl-1-pyrroline-N-oxide; DMSO: dimethylsulfoxide; (-)-ECG: (-)-epicatechin gallate; EGb 761: extract of Ginkgo biloba leaves; ESR spectroscopy: electron spin resonance spectroscopy; GO: glucose oxidase; β-NAD$^+$: β-nicotinamide-adenine dinucleotide, oxidized form; NBT: nitroblue tetrazolium; NP-III: N, N'-bis (2-hydroperoxy-2-methoxyethyl)-1, 4, 5, 8-naphtalene-tetracarboxylic diimide; PAGE: polyacrylamide gel electrophoresis; PEG: polyethylene glycol; PBE: pine bark extract (Pycnogenol®); PBS: phosphate-buffered saline, pH 7.4; SDS: sodium dodecyl sulfate; XDH: xanthine dehydrogenase; XO: xanthine oxidase.

2.3. Superoxide and Hydroxyl Radical Assays

Superoxide anions were generated by UV irradiation of riboflavin/EDTA (Zhao et al., 1989). The reaction mixture contained 0.2 mM riboflavin, 0.4 mM EDTA, and 0.2 M DMPO in the presence and absence of 1 µg/mL of various flavonoids. Hydroxyl radicals were generated by UV photolysis of NP-III and trapped by DMPO (Matsugo et al., 1995). The reaction mixture contained 14 µM NP-III and 0.2 M DMPO, in the presence and absence of 100 µg/mL of various flavonoids. ESR spectra were recorded at room temperature under the conditions as follow: central field 3475 G, modulation frequency 100 kHz, modulation amplitude 2 G, microwave power 10 mW, scan width 200 G, gain 6.3×10^5, temperature 298 K. The scavenging effect of flavonoid on superoxide radical was calculated by $E=[(h_0-h_x)/h_0] \times 100\%$, where h_x and h_0 were the ESR signal intensities of samples with and without flavonoid, respectively.

2.4. Preparation of PBE, EGb 761, and Pure Flavonoids

A stock solution of 10 mg/mL of procyanidins B1, B2, B3, C1, and C2 were prepared in PBS. A stock solution of 100 mg/mL of taxifolin and ECG were prepared in DMSO. A stock solution of 10 mg/mL of PBE, EGb 761, and other pure flavonoids were prepared by dissolving those compounds first in 50 µL of DMSO and then diluting to 1 mL with phosphate buffer, pH 7.4, containing 0.1 mM EDTA.

3. RESULTS

3.1. Effect of PBE on Enzyme Activity

PBE dose-dependently inhibited XO and XDH activities whereas no effect was observed on GO and AO activities (Fig.1).

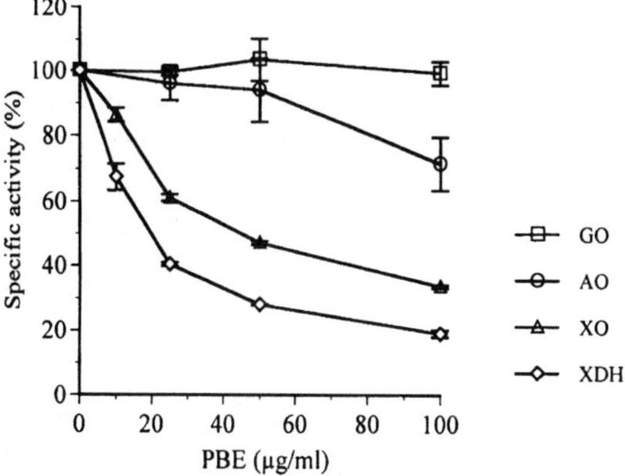

Figure 1. Effect of PBE on enzyme activities. Dose-response curves show the inhibition of xanthine oxidase (XO), and xanthine dehydrogenase (XDH) activities as a function of PBE concentration. Glucose oxidase (GO), and ascorbate oxidase (AO) were not affected by PBE.

Steady-state analysis of XDH activity showed that PBE is an uncompetitive inhibitor of XDH with respect to xanthine as substrate, suggesting that PBE can only bind to the enzyme-substrate complex and inhibit the enzyme activity (Fig. 2A). The plot of the y-axis of Lineweaver-Burk intercept versus PBE concentration (Fig. 2A, insets) defined an apparent K_i of 20.4 µg/mL for XDH. Similarly, PBE exerted an uncompetitive type of inhibition on XO with respect to xanthine as substrate.

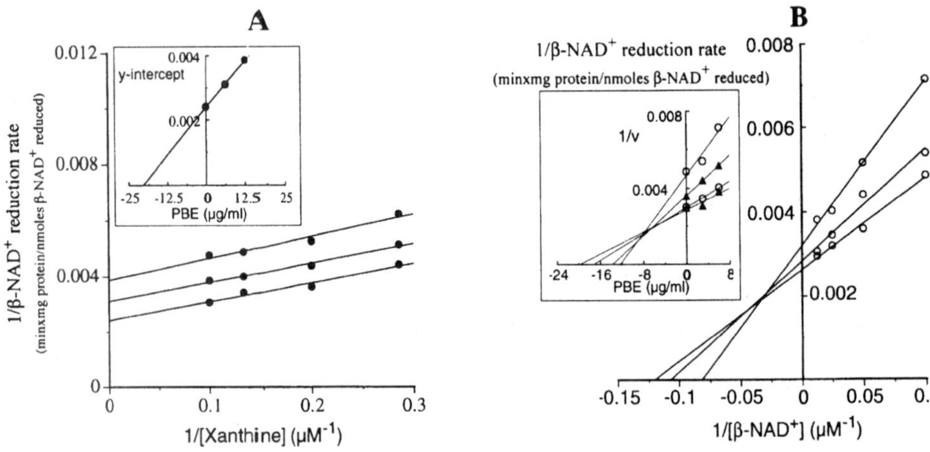

Figure 2. Steady-state analysis of XDH inhibition by PBE. A) β-NAD⁺ concentration was held constant at 500 µM and the concentration of xanthine was varied at a series of fixed concentrations of PBE. Results were presented as Lineweaver-Burk plots and show an uncompetitive type of inhibition. PBE concentrations were (from bottom to top): 0, 6.25, 12.5 µg/mL. The inset is a secondary plot of y-intercept versus PBE concentration. A K_i value of 20.4 µg/mL was defined for the inhibition of XDH by PBE. B) Xanthine concentration was held constant at 150 µM and the concentration of β-NAD⁺ was varied at a series of fixed concentrations of PBE. Results were presented as Lineweaver-Burk plots. PBE concentrations were (from bottom to top): 0, 3, 6 g/mL. Using Dixon plot, a K_i value of 6.9 g/mL was estimated by calculating the median of all intersections (inset). The lines in this plot represent β-NAD⁺ concentrations of (from bottom to top): 80, 40, 20, and 10 µM. The plot shows a mixed type inhibition pattern with respect to β-NAD⁺

It is known that steady state kinetics of both XO and XDH reveals a pattern of parallel lines in the Lineweaver-Burk plots, as one substrate concentration is varied at a series of fixed concentrations of the other substrate, indicating a Ping-Pong steady state mechanism (Olson et al., 1974; Saito and Nishino, 1989). Therefore, the uncompetitive type of inhibition with respect to xanthine could be due to an inhibition in the reduction of the electron acceptors, molecular O_2 and β-NAD⁺. Indeed, steady state experiments with XDH have shown that PBE is a mixed type inhibitor with respect to β-NAD⁺ as substrate (Fig. 2B), indicating that PBE inhibits reduction of β-NAD⁺ to β-NADH that is observed as uncompetitive inhibition with respect to xanthine as substrate. A K_i value of 6.9 µg/mL was estimated from Dixon plot (Fig. 2B, inset) by calculating the median of all intersections. Although the type of inhibition of XO with respect to O_2 by PBE has not been directly assessed, it is expected that a mixed type of inhibition of XO occurs with respect to molecular O_2 by PBE.

3.2. Binding of Proteins by PBE

When subjected to PAGE under non-denaturing conditions, the presence of PBE with XO caused the appearance of a band with a slower electrophoretic mobility , indicating that PBE changes the electrophoretic mobility of XO (Fig. 3). Under non-denaturing conditions, the movement of proteins through the polyacrylamide gel matrix depends on their size, shape, and charge; thus, a change in the electrophoretic mobility of XO in the of presence PBE could be due to the change in one or more of these parameters. Alternatively, the change in the electrophoretic mobility of XO could be due to complexation of XO by PBE. It is important to note that since XO was detected by its activity in the gel, PBE bound XO is enzymatically active and therefore, is not denatured. The presence of SDS (2%) and the reducing agent in the sample buffer did not interfere with the binding of PBE to XO under non-denaturing conditions. Furthermore, the electrophoretic mobility of XO was not changed by PBE under denaturing condition. It appears that binding of PBE to XO is specific for the native structure of the enzyme rather than an unfolded protein structure.

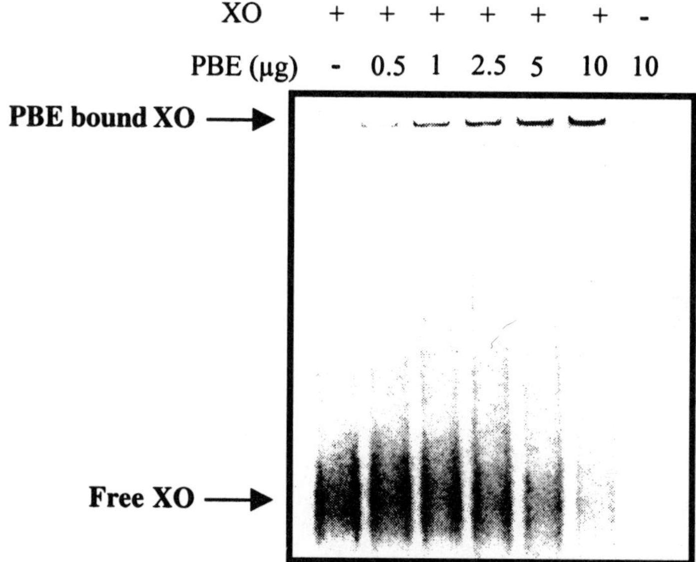

Figure 3. Effect of PBE on the electrophoretic mobility of XO. XO, PBE, and a mixture of XO and PBE at indicated concentrations were subjected to PAGE under non-denaturing condition on a 3-8 % NuPAGE® precast gel. Samples were detected by activity staining. Presence of PBE changed the electrophoretic mobility of XO as revealed by a retarded band in the top part of the gel, indicating that PBE binds to XO.

Binding of PBE to XO and XDH was also confirmed by gel filtration chromatography. The most significant structural difference between XO and XDH is the protein conformation of the 40 Kd domain around FAD. Conformational changes influence the redox potential of the flavin and the reactivity of FAD with the electron acceptors, NAD and molecular oxygen (Saito and Nishino, 1989). Binding of PBE to XDH suggests that the putative binding site(s) for PBE are not affected by the conversion of XO to XDH. However, even a large excess of PBE did not change the electrophoretic mobility of GO under either denaturing or non-denaturing conditions. Although the possible structural differences that cause to the binding of PBE to some proteins but not to others is not known, overall, selectivity is observed for the binding of PBE to the enzymes. By using radioligand-receptor binding assays, it has been shown that procyanidin B3 selectively binds to one receptor and procyanidin B2 binds to two receptors out of 16 receptors tested. Procyanidin B4 inhibits ligand-receptor binding for 5 of the 16 receptors tested, showing less selectivity when compared to procyanidins B2 and B3 (Zhu et al., 1997).

3.3. Type of Interaction Mediates the Binding of PBE to XO

Due to the diversity of molecular structures among procyanidins and the variety of functional groups present in proteins, interactions such as hydrogen bonding, ionic and hydrophobic interactions can take place (Loomis, 1974). When XO and PBE were electro-phoresed in the presence of NaCl or urea, the changed electrophoretic mobility of XO could not be restored. However, Triton X-100 dose-dependently restored the changes in the electrophoretic mobility of XO as observed by the disappearance of the retarded band. This suggests that hydrophobic bonding might be the dominant mode of interaction between PBE and XO (Fig. 4). It has been shown that cytochrome c adsorbed on a column of sepharose containing immobilized polyphenols were effectively eluted by anionic and non-anionic detergents indicating that hydrophobic bonding may be the major mode of interaction between tannin and proteins (Oh et al., 1980; Spencer et al., 1988). Polyethylene glycols (PEGs) are non-detergent competitors of hydrophobic interactions and are extensively used in hydrophobic interaction chromatography (Lee and Lee, 1987; Shibusawa, 1999). PEG 400 has also reversed the inhibition of XO by PBE, confirming the involvement of hydrophobic interactions in the binding of PBE to XO.

3.4. Antioxidant Activity and Inhibitory Effect of Polyphenols on XO

Antioxidant activity of some purified monomeric flavonoids as well as dimer and trimer procyanidins and their effects on XO activity were compared with PBE (Table 1). In terms of superoxide and hydroxyl radical scavenging activity, taxifolin was found to be the most potent antioxidant, whereas an extract from *Ginkgo biloba* leaves (EGb 761) showed the lowest superoxide and hydroxyl radical scavenging activity (Table 1). It is important to note that a minor increase in the production of hydroxyl radical was observed with procyanidins B1, B2, B3 and C1. Pycnogenol® (PBE), epicatechin gallate (ECG) and EGb 761 inhibited XO activity, whereas taxifolin, epicatechin, catechin, and procyanidins C1 and C2 had no effect (Table 1). It has been reported that taxifolin, catechin, and epi-catechin at similar concentrations as used in this study, does not inhibit XO activity (Chang et al., 1993; Cos et al., 1998). The mechanism of inhibition of XO by monomeric flavonoids is not clearly understood. However, it is known that the extent of XO inhibition by struc-turally related flavonoids is quite different (Cos et al., 1998). It is worth mentioning that a small increase in the activity of XO was observed in the presence of procyanidins B1, B2, and B3. However, when compared with their antioxidant activity, no correlation was found between the superoxide/hydroxyl radical scavenging activity of the selected flavonoids and

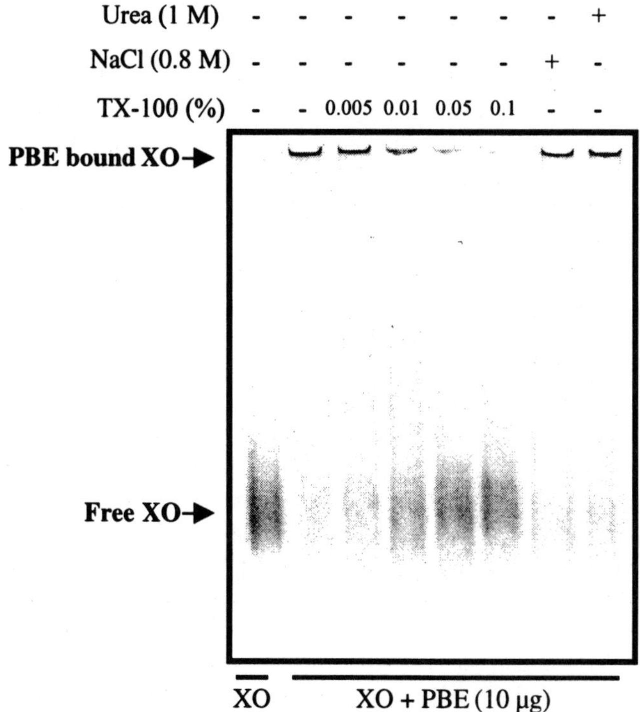

Figure 4. Effects of NaCl, urea, and Triton X-100 on the binding of PBE to XO. XO alone or a mixture of XO and PBE were subjected to non-denaturing PAGE in the presence and absence of indicated concentrations of NaCl, urea, or Triton X-100 (TX-100). Samples were detected by activity staining. The retarded band disappears with increasing concentrations of Triton X-100. Urea, NaCl, or TX-100 alone had no effect on the electrophoretic mobility of XO.

Table 1. Comparative effect of purified and mixture of flavonoids on XO activity and their antioxidant activity. Superoxide radicals were generated in the presence and absence of 1 µg/mL of various flavonoids by UV irradiation of riboflavin/EDTA. Hydroxyl radicals were generated in the presence and absence of 100 µg/mL of various flavonoids by UV photolysis of aqueous solution of NP-III. XO activity was measured in the presence and absence of 50 µg/mL of various flavonoids

Flavonoids	Superoxide scavenged (%)	Hydroxyl radical scavenged (%)	Xanthine oxidase specific activity (%)
Taxifolin	78.5±4.2	41.9±4.1	92.0±6.3
Epicatechin gallate	71.8±4.2	31.4±4.1	49.7±0.4
(-)-Epicatechin	76.6±2.8	18.6±4.7	89.9±2.6
(+)-Catechin	68.0±2.8	16.3±4.7	94.0±0.2
Procyanidin B1	76.6±1.4	-13.7±2.7	112.1±3.1
Procyanidin B2	61.9±2.4	-10.5±4.0	103.9±2.5
Procyanidin B3	52.5±4.9	-8.2±4.7	104.2±3.8
Procyanidin C1	49.2±1.4	-7.4±4.7	98.1±6.4
Procyanidin C2	44.1±4.3	-1.9±1.35	101.0±2.1
PBE	38.0±5.9	2.3±1	47.6±4.6
EGb 761	26.8±2.8	0	63.6±1.5

their inhibitory effect on XO activity. In view of the fact that selected monomer flavonoids as well as dimer and trimer procyanidins were used at the same concentration as PBE, none of these compounds could be primarily responsible for the effect of PBE on XO activity, suggesting that procyanidins with a degree of polymerization of higher than trimer are possibly responsible for PBE action. It has been reported that PBE inhibits angiotensin-converting enzyme (ACE) *in vitro* (Blaszo et al., 1996). Recently PBE was shown to stimulate endothelial nitric oxide synthase (eNOS) activity and increase NO levels in isolated rat aortic ring (Fitzpatrick et al., 1998). The ACE inhibitory and eNOS stimulatory effects of PBE were shown to mostly associate with a fraction of PBE, which contains procyanidins with a higher degree of polymerization than trimer (Blaszo et al., 1996; Fitzpatrick et al., 1998)

3.5. Mechanism of PBE Action on XO

Both PBE and XO are redox active. Upon the binding of xanthine to the molybdopterin containing domain of XO and reduction of Mo(VI), a rapid transfer of electrons occurs from Mo(IV) to the FAD containing domain through other redox sensitive centers of the enzyme (Saito and Nishino, 1989). Therefore, PBE could inhibit XO by either interfering with intramolecular electron transfer processes or by binding to the enzyme and changing its conformation. XO activity may also be affected by a combination of these effects. PAGE experiments have shown that PBE binds to XO and that enzyme inhibition occurs upon binding of PBE to XO, and dissociation of the PBE-XO complex restores enzyme activity. Since no correlation was observed between antioxidant activity and XO inhibition of PBE and other selected purified flavonoids, the results point to the importance of binding of PBE to XO in the enzyme inhibition rather than interference with the intramolecular electron transfer processes. Thereby, in addition to the redox based effects, PBE has direct protein binding properties, adding another biochemical basis of its action in biological systems.

4. REFERENCES

Blaszo, G., Gaspar, R., Gabor, M., Ruve, H. J., and Rohdewald, P., 1996, ACE inhibition and hypotensive effect of a procyanidins containing extract from the bark of *Pinus pinaster* Sol., *Pharm. Pharmacol. Lett.* **1**:8-11.

Chang, W., Lee, Y., Lu, F., and Chiang, H., 1993, Inhibitory effect of flavonoids on xanthine oxidase., *Anticancer Res.* **13**:2165-2170.

Cos, P., Ying, L., Calomme, M., Hu, J. P., Cimanga, K., Van, Poel, B., Pieters, L., Vlietnick, A. J., and Vanden, Berghe, D., 1998, Structure-activity relationship and classification of flavonoids as inhibitors of xanthine oxidase and superoxide scavengers, *J. Nat. Prod.* **61**:71-76.

Elliott, A. J., Scheiber, S. A., Thomas, C., and Pardini, R. S., 1992, Inhibition of glutathione reductase by flavonoids, *Biochem. Pharmacol.* **44**:1603-1608.

Fitzpatrick, D. F., Bing, B., and Rohdewald, P., 1998, Endothelium dependent vascular effects of Pycnogenol[*], *J. Cardiovasc. Pharmacol.* **32**:509-515.

Hagerman, A. E., and , Butler, L. G., 1981, The specificity of proanthocyanidin-protein interaction, *J. Biol. Chem.* **256**:4494-4497.

Hatano, T., Yasuhara, T., Yoshihara, R., Agata, I., Noro, T., and Okuda, T., 1990, Effect of interaction of tannins with co-existing substances: Inhibitory effects of tannins and related polyphenols on xanthine oxidase, *Chem. Pharm. Bull.* **38**:1224-1229.

Kemal, C., Louis-Flamberg, P., Krupinski-Olsen, R., and Shorter, A. L., 1987, Reductive inactivation of soybean lipoxygenase 1 by catechols: A possible mechanism for regulation of lipoxygenase activity, *Biochemistry* **26**:7064-7072.

Kottke, M. K. ., 1998, Scientific and regulatory aspects of nutraceutical products in the United States, *Drug. Dev. Ind. Pharm.* **24**:1177-1195.

Laemmli, U. K. ., 1970, Cleavage of structural proteins during the assembly of the head of bacteriophage, *Nature* **227**:680-685.

Lee, L., and Lee, C., 1987, Thermal stability of proteins in the presence of poly(ethylene glycols), *Biochemistry* **26**:7813-7819.

Loomis, W. D., 1974, Overcoming problems of phenolics and quinones in the isolation of plant enzymes and organelles, *Methods in Enzymol.* **31**:528-544.

Matsugo, S., Yan, L., Han, D., Tritschler, H. J., and Packer, L., 1995, Elucidation of antioxidant activity of dihydrolipoic acid toward hydroxyl radical using a novel hydroxyl radical generator NP-III, *Biochem. Mol. Biol. Inter.* **37**:375-383.

Oh, H., Hoff, J. E., Armstrong, G. S., and Haff, L. A., 1980, Hydrophobic interaction in tannin-protein complexes., *J. Agric. Food Chem.* **28**:394-398.,

Olson, J. S., Ballou, D. P., Palmer, G., and Massey, V., 1974, The reaction of xanthine oxidase with molecular oxygen, *J. Biol. Chem.* **249**:4350-4362.

Ozer, N., Muftuoglu, M., and Ogus, H., 1998, A simple and sensitive method for the activity staining of xanthine oxidase., *J. Biochem. Biophys. Methods* **36**:95-100.

Packer, L., Rimbach, G., and Virgili, F., 1999, Antioxidant activity and biologic properties of a procyanidin-rich extract from pine (*Pinus maritima*) bark, Pycnogenol$^{\text{®}}$, *Free Radic. Biol. Med.* **27**:704-724.

Saito, T., and Nishino, T., 1989, Differences in redox and kinetic properties between NAD-dependent and O_2-dependent types of rat liver xanthine dehydrogenase., *J. Biol. Chem.* **264**:10015-10022.

Shibusawa, Y., 1999, Surface affinity chromatography of human peripheral blood cells, *J. Chromatogr.* B **722**:71-88.

Spencer, C. M., Cai, Y., Martin, R., Gaffney, S. H., Goulding, P. N., Magnolato, D., Lilly, T. H., and Haslam, E., 1988, Polyphenol complexation-some thoughts and observations, *Phytochemistry* **27**:2397-2409.

Yan, Q., and Bennick, A., 1995, Identification of histatins as tannin-binding proteins in human saliva, *Biochem. J.* **311**:341-347.

Zhao, B. L., Li, X. J., He, R., Cheng, S. J., and Xin, W. J., 1989, Scavenging effect of extracts of green tea and natural antioxidants on active radicals, *Cell Biophys.* **14**:175-185.

Zhu, M., Phillipson, J. D., Greengrass, P. M., Bowery, N. E., and Cai, Y., 1997, Plant polyphenols: Biologically active compounds or non-selective binders to proteins? *Phytochemistry* **44**:441-447.

HUMAN 17β-HYDROXYSTEROID DEHYDROGENASE TYPE 5 IS INHIBITED BY DIETARY FLAVONOIDS

A. Krazeisen, R. Breitling, G. Möller, and J. Adamski[1]

1. ABSTRACT

Phytoestrogens contained in a vegetarian diet are supposed to have beneficial effects on the development and progression of a variety of endocrine-related cancers. We have tested the effect of a variety of dietary phytoestrogens, especially flavonoids, on the activity of human 17β-hydroxysteroid dehydrogenase type 5 (17β-HSD 5), a key enzyme in the metabolism of estrogens and androgens. Our studies show that reductive and oxidative activity of the enzyme are inhibited by many compounds, especially zearalenone, coumestrol, quercetin and biochanin A. Among flavones, inhibitor potency is enhanced with increased degree of hydroxylation. The most effective inhibitors seem to bind to the hydrophilic cofactor binding pocket of the enzyme.

2. INTRODUCTION

Epidemiological studies have shown that environmental factors, especially dietary compounds, have a major impact on cancer development and progression. One class of substances suggested to be responsible for these cancer protective effects are the phytoestrogens which are particularly abundant in soy products, a major component of the Asian diet. Consequently, incidence of breast and prostate cancer is significantly lower in Asia compared to Northern Europe and America (Flanders, 1984; Hirayama, 1979; Shimizu et al., 1991). Soy products have been shown to possess chemopreventive effects in experimentally induced cancers (Barnes et al., 1990; Makela et al., 1991).

---

[1] Corresponding author: J. Adamski, GSF-National Research Center for Environment and Health, Institute for Experimental Genetics, Genome Analysis Center, Ingolstädter Landstraße 1, 85764 Neuherberg, Germany, Tel.: +49-89-3187-3155, FAX: +49-89-3187-3225; e-mail: adamski@gsf.de

Flavonoids in Cell Function, edited by Béla Buslig and John Manthey.
Kluwer Academic/Plenum Publishers, New York, 2002.

Phytoestrogens are plant-derived, non-steroidal compounds possessing estrogenic activity. They can be divided into three main classes: flavonoids, coumestans and lignans. All of these are diphenolic substances with structural similarity to natural human steroid hormones. Our study focused on the group of flavonoids (flavones, flavanones and isoflavones), which are found in almost all plant families in leaves, stems, roots, flowers and seeds (Harborne, 1971; Verdeal and Ryan, 1979), where they act as natural fungicides, UV-protectants, and flower pigments (Harborne, 1971).

The soybean is the main dietary source for isoflavones. Flavones and flavanones are widely distributed in all plant families and are found in fruits, vegetables, berries, herbs, beans, and green tea. High levels of coumestrol are found in alfalfa and various beans (Strauss et al., 1998). Plant lignans are found in fibre rich-food, such as unrefined grain products, and are transformed by intestinal bacteria into mammalian lignans such as enterolactone and enterodiol (Kurzer and Xu, 1997). Additional dietary estrogenic compounds we tested are the resorcylic acid lactones, e.g. zearalenone, a secondary metabolites in mold infected food, and the saponin 18β-glycyrrhetinic acid, a major component of licorice.

Phytoestrogens are thought to act as chemopreventive agents via changing hormone concentrations, mainly by mimicking endogenous hormones either at the hormone receptor or at key enzymes of hormone metabolism, thus affecting the level of active steroids. They may also have diverse non-hormonal effects (reviewed in Kurzer and Xu, 1997). 17β-hydroxysteroid dehydrogenases play an essential role in the formation or inactivation of active steroid hormones controlling the last step in the production of all androgens and estrogens via reduction or oxidation of steroids at position 17. Nine types of 17β-hHSDs have been described to date varying in substrate specificity and tissue distribution (Peltoketo et al., 1999).

In contrast to all the other 17β-hHSDs, 17β-hHSD 5 is a member of the aldo-keto (AKR) reductase family. This protein family is characterized by a typical conserved catalytic tetrad of amino acids (Tyr55, His84, His117, Asp50), which participates in a "proton-relay" mechanism, with Tyr55 representing the major acid/base catalyst (Penning, 1999; Schlegel et al., 1998).

Multifunctionality is a special characteristic of human 17β-HSD 5 which enables conversion at the 3α-, 17β- and the 20α-position of estrogens, androgens as well as progestins (Dufort et al., 1999). In humans, 17β-HSD type 5 is expressed in reproductive and hormone target tissues (El-Alfy et al., 1999; Pelletier et al., 1999), e.g. in ovary, uterine endometrium, mammary gland, testis, and adrenal gland. Additionally, the enzyme has been detected in prostate indicating the possibility of local steroidogenesis from circulating inactive precursors (Fig. 1). Substrate specificity of 17β-hHSD type 5 is comparable to 17β-HSD type 3 which catalyzes the conversion of androstenedione to testosterone (Dufort et al., 1999). It also degrades the active androgen 5α-DHT to androstanediol and subsequently to androsterone in the prostate (Lin et al., 1997). Consequently it was suggested that the enzyme controls the occupancy of the androgen receptor and regulates local androgen concentration (Lin et al., 1997), which is instrumental for the control of normal and abnormal growth of the prostate.

In this work we determined the inhibition pattern of human 17β-HSD 5, a key enzyme of androgen and estrogen metabolism, by flavonoids and other dietary phytoestrogens, and evaluated which structural features might be responsible for inhibitory potency.

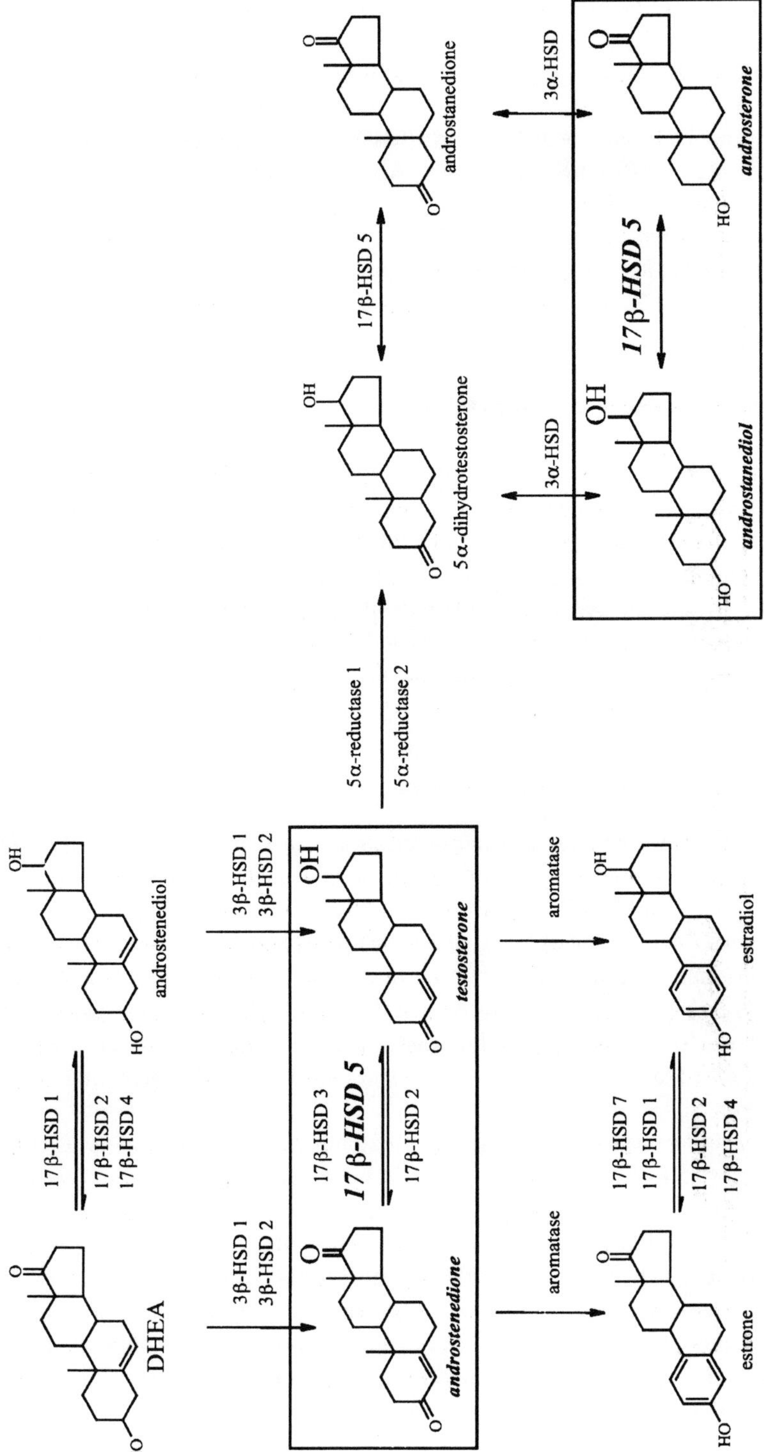

Fig. 1. Intracrinological role of 17β-hHSD 5. Unique among animal species, in humans or other primates the adrenals secrete large amounts of inactive precursor steroids (DHEA, DHEA-S, and androstenedione), which are converted *in situ* into potent androgens and estrogens in peripheral tissues by many different HSDs (Krazeisen et al., 1999; Labrie et al., 1997). Tested 17β-HSD activities of 17β-hHSD 5 are shown in boxes.

3. MATERIALS AND METHODS

3.1. Phytoestrogens, Related Compounds, Cofactors and Substrates

Abietic acid, quercetin and kaempferol were purchased from ICN Biochemicals GmbH. Coumestrol was obtained from Steraloids Inc. All other inhibitors were obtained from Sigma Aldrich Chemie GmbH at the maximum purity available. The cofactors NAD (Serva) and NADP (from Sigma Aldrich Chemie GmbH) were used for oxidation and reduction, respectively. The radioactive substrates androst-4-ene-3,17-dione [1,2,6,7-^{3}H(N)] and 5α-androstane-3α-17β-diol [9,11-^{3}H(N)] were obtained from NEN-Life Science Products Inc.

3.2. Expression of Recombinant 17β-hHSD 5

For protein expression the complete coding sequence of 17β-hHSD 5 (kindly provided by T. Penning) was cloned into the *BamH I* and *Kpn I* sites of *pGEX* as described (Leenders et al., 1996). Transformed bacteria (*E. coli* strain JM 107) containing the pGEX-17β-hHSD 5 plasmid were grown in LB-media containing 50μg/ml ampicillin at 37°C. Expression of GST-fusion protein was induced at an absorbance of A_{600nm} = 0.6 by addition of 0.25mM IPTG. 4h after induction bacteria were harvested and the protein was purified as described (Leenders et al., 1996). 17β-hHSD 5 released from the fusion protein by thrombin cleavage was used for the enzymatic assays (Möller et al., 1999).

3.3. Enzyme Assays

Reductive 17β-hHSD activity was measured in a final volume of 500μl of 50mM phosphate buffer, pH 7.4, including 8μg of protein, androstenedione (androst-4-ene-3,17-dione [1,2,6,7-^{3}H(N)], 30nM) with NADPH (final 600μM) as cofactor. Oxidative activity was determined in a final volume of 500μl of 100mM phosphate buffer, pH 9.0 using 6μg of enzyme, androstanediol (5α-androstane-3α-17β-diol [9,11-^{3}H(N)], 50nM) with NAD (final 730μM) as cofactor. Phytoestrogens diluted in ethanol or DMSO were added to the reaction mixture to a final concentration between 20μM and 100nM. Turbidity measurements were performed to determine the maximum solubility of the different inhibitors in final assay solution.

The reaction mixture was incubated at 37°C for 20min. *In vitro* conditions as described were chosen to ensure that conversion was still in the linear range. In control reactions (without inhibitors) about 30% of the substrate was converted into product within incubation time (10 - 20nmol). Results are shown for two independent expression experiments. The reaction was stopped by the addition of ascorbate/ethanol . The steroids were prepurified on reversed phase (C18) columns. Prepurified steroids were separated on an RP18 column using high performance liquid chromatography with the mobile phase of acetonitrile/water (1:1 v/v) as previously described (Adamski et al., 1989). Radioactive compounds were monitored by on-line scintillation counting.

4. RESULTS AND DISCUSSION

4.1. Inhibition of the Human 17β-hHSD 5 by Phytohormones

Several dietary phytoestrogens were tested for their effect on 17β-hHSD 5. The panel of tested substances included flavonoids with different hydroxylation patterns as well as coumarin, coumestrol, and other dietary compounds of known estrogenic potential. Results of inhibition studies for two different metabolic pathways catalyzed by 17β-hHSD 5 are shown in Tables 1 and 2. We tested final inhibitor concentrations of 100nM, 1μM, 5μM, 10μM and 20μM.

Table1. Inhibition of conversion of androstenedione to testosterone. Results of *in vitro* assays measuring the influence of environmental hormones on reductive activity of 17β-hHSD 5. Inhibition is shown in % at all 5 concentrations of phytoestrogen tested. IC_{50} values of dietary phytohormones for reduction of androstenedione to testosterone as obtained by graphical determination

	Inhibition [%]					IC_{50} [μM]
	20μM	10μM	5μM	1μM	100nM	
flavanone	7	6	11	0	7	50
naringenin	37	21	16	2	0	33
flavone	22	15	14	14	8	41
3-hydroxyflavone	40	24	25	3	5	>20*
5-hydroxyflavone	31	19	13	11	8	>20*
7-hydroxyflavone	46	32	23	10	0	24
3,7-dihydroxyflavone	45	28	25	11	2	20
chrysin	50	35	21	5	8	20
apigenin	54	33	24	8	5	20
kaempferol	50	30	18	0	0	20
quercetin	61	53	41	10	0	9
5-methoxyflavone	41	0	2	4	0	50
daidzein	19	2	3	0	0	>20
genistein	49	20	15	11	0	>20
biochanin A	61	44	26	7	0	14
coumarin	7	12	0	2	5	>20
coumestrol	78	64	49	4	0	5
abietic acid	49	38	21	2	0	20
18β-glycyrrhetinic acid	59	52	32	10	5	30
DES	39	30	14	8	3	>20
tamoxifen	24	20	10	2	0	>20
zearalenone	74	58	57	15	0	4

*not soluble at concentrations higher than 20μM

Our experiments substantiate that a major part of the tested dietary hormones has inhibitory effects on both, the reductive and oxidative, activities of 17β-hHSD 5 analyzed here. Reduction of androstenedione to testosterone as well as oxidation of androstanediol to androsterone are affected. The most potent inhibitors are coumestrol, zearalenone, the flavone quercetin and the isoflavone biochanin A (the precursor of genistein). IC_{50} values are also listed in Tables 1 and 2.

Table 2. Inhibition of conversion of androstanediol to androsterone. Results of *in vitro* assays measuring the influence of environmental hormones on oxidative activity of 17β-hHSD 5. Inhibition in % is shown at all 5 concentrations of phytoestrogen tested. IC_{50} value determination as described above

	Inhibition [%]					IC_{50} [μM]
	20μM	10μM	5μM	1μM	100nM	
flavanone	17	8	5	0	0	>20
naringenin	59	49	25	0	0	10
flavone	3	0	2	0	0	>20*
3-hydroxyflavone	6	3	3	0	0	20
5-hydroxyflavone	25	22	0	0	0	>20*
7-hydroxyflavone	72	54	42	0	8	7
3,7-dihydroxyflavone	54	38	31	13	0	18
chrysin	55	44	15	7	4	13
apigenin	35	25	3	0	0	45
kaempferol	58	57	37	0	0	8
quercetin	78	71	52	17	4	5
5-methoxyflavone	27	20	17	0	0	>20
daidzein	19	7	2	1	0	>20
genistein	31	15	5	0	0	>20
biochanin A	68	54	36	2	3	8
coumarin	28	25	0	0	5	>20
coumestrol	58	37	42	36	34	11
abietic acid	57	48	26	3	16	10
18β-glycyrrhetinic acid	8	3	0	0	2	>20
DES	48	31	13	0	0	20
tamoxifen	10	8	7	0	6	>20
zearalenone	79	71	68	29	23	2

*not soluble at concentrations higher than 20μM

On the other hand, substances like the synthetic antiestrogen tamoxifen, the strong natural estrogens daidzein and coumarin had no influence on both types of reaction. 18β-glycyrrhetinic acid strongly inhibited the reduction of androstenedione but had no effect on the oxidative reaction.

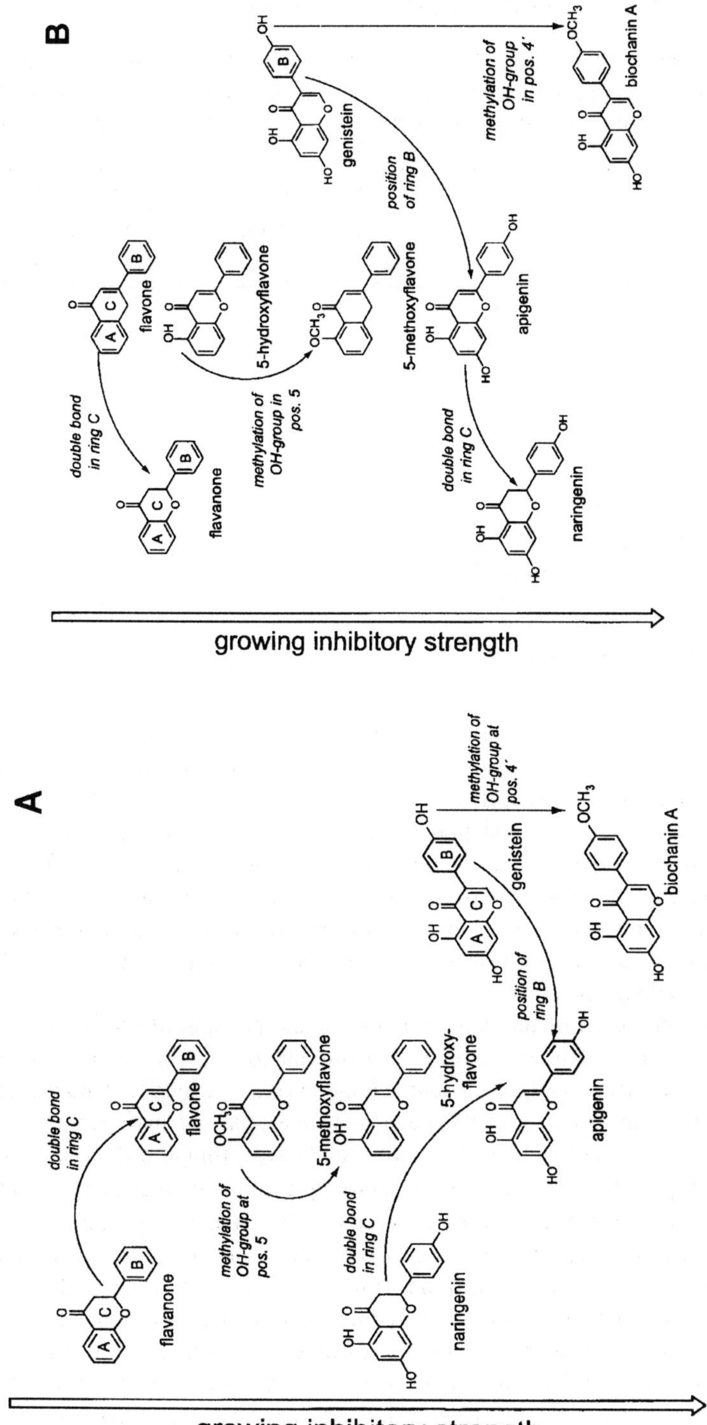

Fig. 2. Inhibitory influence of dietary components A) on the reduction of androstenedione to testosterone B) on the oxidation of androstanediol to androsterone. All substances are listed in order of their inhibitor potency. Within each row inhibitory capacity is increasing from top to bottom.

4.2. Inhibitory Flavonoids and Their Structural Requirements

Table 3. Structure-dependent inhibition by flavones. To evaluate the importance of single structural elements we compared flavones with different hydroxylation patterns. Inhibition for the reductive and oxidative reactions are presented in % at a final concentration of 20μM of inhibitors.

	Inhibition in % (inhibitor conc. 20μM)	
	androstenedione to testosterone	androstanediol to androsterone
flavone	22	3
3-hydroxyflavone	40	6
5-hydroxyflavone	31	25
5-methoxyflavone	41	27
3,7-dihydroxyflavone	45	54
4',5,7-trihydroxyflavone (apigenin)	54	55
5,7-dihydroxyflavone (chrysin)	50	55
3,4',5,7-tetrahydroxyflavone (kaempferol)	50	58
7-hydroxyflavone	46	72
3,3',4',5,7-pentahydroxyflavone (quercetin)	61	78

To analyze the importance of structural elements within the large group of flavonoids we compared inhibitory potency of flavones in dependence on hydroxylation (Table 3), and compared inhibitory action of a large variety of flavonoids (Fig. 2A, 2B). For both direction of conversions, increasing hydroxylation of the flavone nuclei increases inhibitory activity. This is supported by the observation that the non-estrogenic components flavone and flavanone (Miksicek, 1995) show almost no inhibitory activity. This observation indicates that the hydrophilic character of the plant components is necessary for inhibitor potency and possibly directs this substrate towards more polar elements of the enzyme, e. g. the hydrophilic cofactor binding pocket.

To determine the influence of the double bonds, or the position of the B ring, we compared inhibitory strength between members showing identical hydroxylation pattern (Fig. 2A, 2B). Under reductive conditions a double bond in ring C, which is characteristic for flavones, increases the inhibitory strength (flavanone versus flavone, naringenin versus apigenin). In contrast, flavanones lacking the double bond have stronger influence on the oxidative activity of 17β-hHSD 5 (flavone *vs* flavanone, apigenin *vs* naringenin). For both 17β-hHSD activities inhibitor potency decreases when ring B is found at position 3 as it is in isoflavones. Flavones carrying a 5-methoxy group on the A ring are more potent than substances with a hydroxy group at this position. However, within the group of flavones methylation of position 5 increases inhibition of the reductive pathway but decreases activity in oxidative conversion. Within the group of isoflavones methylation of hydroxy group at position 4' causes high inhibitory activity. Our *in vitro* assays have shown that reductive and oxidative activity of 17β-hHSD 5 are influenced by almost the same substances. Zearalenone, quercetin, biochanin A and coumestrol are the strongest inhibitors.

4.3. Postulated Inhibitory Mechanism

The exact orientation of the inhibitors cannot be determined from our data. However, increased inhibitory potency with an increasing degree of hydroxylation of flavonoids indicates that these substances exert their inhibitory effect by binding to the hydrophilic cofactor binding site and not to the apolar substrate binding cleft.

In the ternary complex of 3α-HSD, NADP and testosterone the 3α-hydroxy group of the substrate is situated between the conserved tyrosine and histidine of the active center and forms hydrogen bonds with both residues (Bennett et al., 1997). The same position is occupied by a tightly bound water in the binary complex of 3α-HSD (Bennett et al., 1997) and by the oxygen atom of several inhibitors with other members of the AKR family. The importance of hydroxylation at position 7 (this work and Le Bail et al., 1998) of the flavonoids suggests that this hydroxy group could also bind to the active center.

4.4. Main Dietary Sources of Phytoestrogens and Plasma Concentrations

The most potent inhibitors are present in very common dietary products, with the exception of zearalenone, which is found in mold infected food. Quercetin (a flavone) is found in e.g. apples, onions, chamomile, and tea. Biochanin A (an isoflavone) as well as the strong inhibitor coumestrol are in a variety of beans, especially soybeans, a major component of the Asian diet. Consequently, high plasma values of isoflavones have been observed in Japanese man (Adlercreutz et al., 1993; Adlercreutz et al., 1994). The highest individual value exceeded 2μM, compared to a mean plasma value for daidzein and genistein in Finnish subjects of about 5nM each (Adlercreutz, 1998). These high local phytoestrogen levels might alter local steroid hormone concentration, e.g. by inhibiting 17β-hHSD 5, as shown in the present study.

The involvement of 17β-hHSD 5 inhibition in delay of breast cancer development is not clear. Breast cancer is associated with high local estrogen concentrations. This steroid hormone can be produced locally from estrone by 17β-HSD 1 which has been shown to be inhibited by genistein and coumestrol (Makela et al., 1995). Other sources of estradiol would be production from testosterone which is built by 17β-HSD 5 and converted by aromatase. To some extend 17β-HSD 5 converts estrone to estradiol (Dufort et al., 1999). Consequently an inhibition of this enzyme might effect the local production of active estrogens and thus influence breast cancer development.

One interesting effect is seen in the inhibitory capacity of 18β-glycyrrhetinic acid (Tables 1 and 2). This substance found in licorice does not influence the oxidative pathway but inhibits reduction of androstenedione to testosterone. Previously it has been observed that the serum testosterone level is significantly reduced in men consuming about 7g of a commercial preparation of licorice (containing 0.5g of 18β-glycyrrhetinic acid) a day. It has been demonstrated that licorice consumption inhibits 11β-hHSD (Walker et al., 1995), and 17β-hHSD and 17,20-lyase activity (Armanini et al., 1999). Reduced 17β-HSD activity might be due to inhibiton of 17β-hHSD type 3 or 5. Decreased testosterone levels result in reduced libido or other sexual dysfunction but might have beneficial effects in cases of abnormal prostate growth.

We conclude that 17β-HSD 5 is a potential target for the inhibitory effect of a variety of phytoestrogens. This inhibition might contribute significantly to the cancer preventive action of a soy-based diet. The determination of structural requirements for the inhibitory effects may be a first step towards a rational design of anti-17β-HSD 5 drugs, to be employed in the treatment or prevention of hormone-related cancers.

5. ACKNOWLEDGMENTS

This work was supported in part by a DFG grant to J. Adamski. We thank T. Penning for kindly providing the pET-plasmid coding for 17β-HSD 5.

7. REFERENCES

Adamski, J., Sierralta, W. D., and Jungblut, P. W., 1989, Assignment of estradiol-17β-dehydrogenase and of estrone reductase to cytoplasmic structures of porcine endometrium cells, *Acta Endocrinol.* (Copenh.). **121**:161-167.

Adlercreutz, H., 1998, Epidemiology of phytoestrogens, Bailliere's *Clin. Endocrinol. Metab.* **12**:605-623.

Adlercreutz, H., Fotsis, T., Watanabe, S., Lampe, J., Wahala, K., Makela, T., and Hase, T., 1994, Determination of lignans and isoflavonoids in plasma by isotope dilution gas chromatography-mass spectrometry, *Cancer Detect. Prev.* **18**:259-271.

Adlercreutz, H., Markkanen, H., and Watanabe, S., 1993, Plasma concentrations of phyto-oestrogens in Japanese men, *Lancet* **342**:1209-1210.

Armanini, D., Bonanni, G., and Palermo, M., 1999, Reduction of serum testosterone in men by licorice [letter], *New Engl. J. Med.* **341**:1158.

Barnes, S., Grubbs, C., Setchell, K. D., and Carlson, J., 1990, Soybeans inhibit mammary tumors in models of breast cancer, *Prog. Clin. Biol. Res.* **347**:239-253.

Bennett, M. J., Albert, R. H., Jez, J. M., Ma, H., Penning, T. M., and Lewis, M., 1997, Steroid recognition and regulation of hormone action: crystal structure of testosterone and NADP+ bound to 3α-hydroxysteroid/dihydrodiol dehydrogenase, *Structure* **5**:799-812.

Dufort, I., Rheault, P., Huang, X. F., Soucy, P., and Luu-The, V., 1999, Characteristics of a highly labile human type 5 17β-hydroxysteroid dehydrogenase, *Endocrinology* **140**:568-574.

El-Alfy, M., Luu-The, V., Huang, X. F., Berger, L., Labrie, F., and Pelletier, G., 1999, Localization of type 5 17β-hydroxysteroid dehydrogenase, 3β-hydroxysteroid dehydrogenase, and androgen receptor in the human prostate by in situ hybridization and immunocytochemistry, *Endocrinology* **140**:1481-1491.

Flanders, W. D., 1984, Review: Prostate Cancer Epidemiology, *Prostate* **5**:621-629.

Harborne, J., 1971, Distribution of flavonoids in leguminosae, in: *Chemotaxonomy of the Leguminosae,* Harborne, J. B., Boulter, D. and Turner, B. L., Eds. Academic Press, London, pp. 31-71.

Hirayama, T., 1979, Epidemiology of prostate cancer with special reference to the role of diet, *Natl. Cancer Inst. Monogr.* 149-155.

Krazeisen, A., Breitling, R., Imai, K., Fritz, S., Möller, G., and Adamski, J., 1999, Determination of cDNA, gene structure and chromosomal localization of the novel human 17β-hydroxysteroid dehydrogenase type 7, *FEBS Lett.*. **460**:373-379.

Kurzer, M. S., and Xu, X., 1997, Dietary phytoestrogens, Annu. Rev. Nutr. **17**:353-381.

Labrie, F., Luu-The, V., Lin, S. X., Labrie, C., Simard, J., Breton, R., and Belanger, A., 1997, The key role of 17β-hydroxysteroid dehydrogenases in sex steroid biology, *Steroids* **62**:148-158.

Le Bail, J. C., Laroche, T., Marre-Fournier, F., and Habrioux, G., 1998, Aromatase and 17β-hydroxysteroid dehydrogenase inhibition by flavonoids. *Cancer Lett.* **133**:101-106.

Leenders, F., Tesdorpf, J. G., Markus, M., Engel, T., Seedorf, U., and Adamski, J., 1996, Porcine 80-kDa protein reveals intrinsic 17β-hydroxysteroid dehydrogenase, fatty acyl-CoA-hydratase/dehydrogenase, and sterol transfer activities. *J. Biol. Chem.* **271**:5438-5442.

Lin, H. K., Jez, J. M., Schlegel, B. P., Peehl, D. M., Pachter, J. A., and Penning, T. M., 1997, Expression and characterization of recombinant type 2 3α-hydroxysteroid dehydrogenase (HSD) from human prostate: demonstration of bifunctional 3α/17β-HSD activity and cellular distribution [published erratum appears in *Mol. Endocrinol.* 1999, Nov. **12**(11):1763]. *Mol. Endocrinol.* **11**:1971-1984.

Makela, S., Poutanen, M., Lehtimaki, J., Kostian, M. L., Santti, R., and Vihko, R., 1995, Estrogen-specific 17β-hydroxysteroid oxidoreductase type 1 (E.C. 1.1.1.62) as a possible target for the action of phytoestrogens. *Proc. Soc. Exp. Biol. Med.* **208**:51-59.

Makela, S., Pylkkänen, L., Santti, R., and Adlercreutz, H., 1991, Role of plant estrogens in normal and estrogen-related altered growth of the mouse prostate. EURO FOOD TOX III, *Proceedings of the Interdisciplinary Conference on Effects of Food on Immune and Hormonal Systems*, Zürich, Switzerland, pp. 135-139.

Miksicek, R. J., 1995, Estrogenic flavonoids: structural requirements for biological activity. *Proc. Soc. Exp. Biol. Med.* **208**:44-50.

Möller, G., Luders, J., Markus, M., Husen, B., Van Veldhoven, P. P., and Adamski, J., 1999, Peroxisome targeting of porcine 17β-hydroxysteroid dehydrogenase type IV/D-specific multifunctional protein 2 is mediated by its C-terminal tripeptide AKI. *J. Cell Biochem.* **73**:70-78.

Pelletier, G., Luu-The, V., Tetu, B., and Labrie, F., 1999, Immunocytochemical localization of type 5 17β-hydroxysteroid dehydrogenase in human reproductive tissues. *J. Histochem. Cytochem.* **47**:731-738.

Peltoketo, H., Luu-The, V., Simard, J., and Adamski, J., 1999., 17β-hydroxysteroid dehydrogenase (HSD)/17-ketosteroid reductase (KSR) family; nomenclature and main characteristics of the 17HSD/KSR enzymes. *J. Mol. Endocrinol.* **23**:1-11.

Penning, T. M., 1999, Molecular determinants of steroid recognition and catalysis in aldo-keto reductases. Lessons from 3α-hydroxysteroid dehydrogenase. *J. Steroid Biochem. Mol. Biol.* **69**:211-225.

Schlegel, B. P., Jez, J. M., and Penning, T. M., 1998, Mutagenesis of 3α-hydroxysteroid dehydrogenase reveals a "push-pull" mechanism for proton transfer in aldo-keto reductases. *Biochemistry* **37**:3538-3548.

Shimizu, H., Ross, R. K., Bernstein, L., Yatani, R., Henderson, B. E., and Mack, T. M., 1991, Cancers of the prostate and breast among Japanese and white immigrants in Los Angeles County. *Brit. J. Cancer* **63**:963-966.

Strauss, L., Santti, R., Saarinen, N., Streng, T., Joshi, S., and Makela, S., 1998, Dietary phytoestrogens and their role in hormonally dependent disease. *Toxicol. Lett.* **102-103**:349-354.

Verdeal, K., and Ryan, D., 1979, Naturally occurring estrogens in plant foodstuffs - A review. *J. Food Protection* **42**:557-583.

Walker, B. R., Aggarwal, I., Stewart, P. M., Padfield, P. L., and Edwards, C. R., 1995, Endogenous inhibitors of 11β-hydroxysteroid dehydrogenase in hypertension. *J. Clin. Endocrinol. Metab.* **80**:529-533.

INTERACTIONS OF FLAVONES AND OTHER PHYTOCHEMICALS WITH ADENOSINE RECEPTORS

Kenneth A. Jacobson[1]*, Stefano Moro[1,2], John A. Manthey[3], Patrick L. West[1], and Xiao-duo Ji[1]

1. ABSTRACT

Dietary flavonoids have varied effects on animal cells, such as inhibition of platelet binding and aggregation, inhibition of inflammation, and anticancer properties, but the mechanisms of these effects remain largely unexplained. Adenosine receptors are involved in the homeostasis of the immune, cardiovascular, and central nervous systems, and adenosine agonists/antagonists exert many similar effects. The affinity of flavonoids and other phytochemicals to adenosine receptors suggests that a wide range of natural substances in the diet may potentially block the effects of endogenous adenosine. We used competitive radioligand binding assays to screen flavonoid libraries for affinity and a computational CoMFA analysis of flavonoids to compare steric and electrostatic requirements for ligand recognition at three subtypes of adenosine receptors. Flavone derivatives, such as galangin, were found to bind to three subtypes of adenosine receptors in the μM range. Pentamethylmorin (K_i 2.65 μM) was 14- to 17-fold selective for human A_3 receptors than for A_1 and A_{2A} receptors. An isoflavone, genistein, was found to bind to A_1 receptors. Aurones, such as hispidol (K_i 350 nM) are selective A_1 receptor antagonists, and, like genistein, are present in soy. The flavones, chemically optimized for receptor binding, have led to the antagonist, MRS 1067 (3,6-dichloro-2'-(isopropoxy)-4'-methylflavone), which is 200-fold more selective for human A_3 than A_1 receptors. Adenosine receptor antagonism,

[1] Molecular Recognition Section, Laboratory of Bioorganic Chemistry, National Institute of Diabetes, Digestive and Kidney Diseases, National Institutes of Health, Bethesda, MD 20892-0810. [2] Pharmaceutical Science Department, University of Padova, Italy. [3] U.S. Department of Agriculture, Agricultural Research Service, U.S. Citrus and Subtropical Products Laboratory, P.O. Box 1909, Winter Haven, FL 33883; *Correspondence to: Dr. K. A. Jacobson, Chief, Molecular Recognition Section, LBC, NIDDK, NIH, Bldg. 8A, Room B1A-19, Bethesda, MD 20892-0810 Tel.: (301) 496-9024, FAX: (301) 480-8422. e-mail: kajacobs@helix.nih.gov

Flavonoids in Cell Function, edited by Béla Buslig and John Manthey.
Kluwer Academic/Plenum Publishers, New York, 2002.

therefore, may be important in the spectrum of biological activities reported for the flavonoids.

2. INTRODUCTION

Flavonoids are ubiquitous phenolic compounds occurring throughout the plant kingdom. Ingested flavonoids have recently been recognized to have many surprising and exciting implications when taken as dietary supplements. Flavonoids have been linked to many beneficial effects in humans, such as inhibition of platelet binding and aggregation (for review see Beretz and Cazenave, 1988), inhibition of inflammation (for a review see Middleton and Ferriola, 1988; Manthey, 2000), and anticancer properties (for a review see Wang et al., 1998). Some flavonoids also have estrogen-like effects and/or inhibit tyrosine kinases (Barnes et al., 1999). The mechanisms of many of these effects, however, remain largely unexplained (Karton et al., 1996).

Flavonoids from plant sources comprise an important component of the human diet. Concentrations of these compounds are especially high in legumes, such as soybeans. Soy proteins are included in many foods and have been found to contain concentrations of 0.1-3.0 mg of flavonoids per gram (Coward et al., 1993). Data indicates that consuming "modest" quantities of these flavonoids can result in high levels in the circulatory system (Setchell, 1996).

Adenosine receptors are involved in the homeostasis of the immune, cardiovascular, and central nervous systems. Activation of adenosine receptors is associated with cerebro-protective (von Lubitz et al., 1994) and cardioprotective (Stickler et al., 1996) properties, and also with effects on the immune and inflammatory systems (Sajjadi et al., 1996). A_3 receptor-selective antagonists have been proposed to have anti-asthmatic (Beaven et al., 1994) and possibly cerebroprotective (Jacobson et al., 1995) properties. We have noted that some of the observed effects for flavonoids are similar to *in vivo* effects related to activation or antagonism of adenosine receptors. Therefore, we have explored the interactions between purine receptors and certain flavonoids as a possible mechanism for the observed effects of flavonoids in humans.

A wide variety of non-purine ligands that bind selectively to adenosine receptors have been described (Jacobson et al., 1997; Müller, 1997). The availability of selective ligands has facilitated studies of the physiological roles of particular subtypes of adenosine receptors. The A_3 receptors have already been implicated in vascular effects, inflammation and cancer, *i.e.* three areas in which flavonoids have been considered biologically active. One example is the common effect that both flavonoids and adenosine have on histamine release.

A broad screening of phytochemicals, using competition for specific radioligand binding to human A_3 receptors as an assay, demonstrated moderate affinities, with K_i values in the micromolar range (Moro et al., 1998). This suggests, that if blood levels reach this range following ingestion of flavonoids, the flavonoids may antagonize the activity of these receptors (Karton et al., 1996; Ji et al., 1995). These findings might ultimately help to elucidate a mechanism for the effects that have been attributed to ingested flavonoids, and warrant further screening of flavonoids.

3. RESULTS AND LITERATURE REVIEW

3.1. Flavonoids in Chemical Library Identified as Adenosine Antagonists

The primary screening assay consisted of single point competition at fixed flavonoid concentration (10 μM) for specific binding of ^{125}I-AB-MECA by recombinant human A_3 receptors expressed in HEK-293 cells, and promising candidates were further examined for concentration dependence in binding (Ji et al., 1995).

Flavone derivatives, such as galangin, were found to bind to three subtypes of adenosine receptors in the μM range. Galangin displayed K_i values of 1 μM for both rat A_1 and A_{2A} receptors and 3 μM for human A_3 receptors. Methylation but not acetylation of the hydroxyl groups of galangin enhanced A_3 affinity. Triethyl- and tripropyl-galangin displayed K_i values of 0.3-0.4 μM at human A_3 receptors (Karton et al., 1996). Pentamethylmorin showed 14- to 17-fold selectivity for human A_3 receptors vs. rat A_1 and A_{2A} receptors, with a K_i value of 2.65 μM. Several polymethoxylated flavones from citrus, tetramethylscutellarein and tetramethyl-kaempferol, also showed high affinities for the A_1 receptor. Reduction of the 2,3-olefinic bond, as in (±)-dihydroquercetin, or glycosidation, as in robinin, greatly diminished affinity. α-Naphthoflavone had greater receptor affinity (0.79 μM to A_1 receptors) than the β-isomer. Other natural products of plant origin, including oxogalanthine lactam, hematoxylin, and arborinine were found to bind to A_1 adenosine receptors with K_i values of 3-13 μM. These findings indicate that flavonoids and other phytochemicals may provide leads for the development of novel adenosine antagonists.

A naturally occurring isoflavone, genistein, was found to bind with moderate affinity to A_1 receptors (Okajima et al., 1994), but only very weakly to A_3 receptors (Ji et al., 1995). We have examined the affinity of other isoflavones to A_1 and A_{2A} receptors and found that genistein is the most potent (Table 1).

The principle of adenosine antagonism as a mechanism for the observed biological effects of flavonoids has been demonstrated. Cirsimarin and cirsimaritin, flavonoids of *Microtea debilis* (Phytolaccaceae), were found to have adenosine antagonistic properties in rats (Hasrat, 1997). In traditional medicine *Microtea debilis* is used against proteinuria. Cirsimarin was isolated as the active component and was shown to be an A_1 adenosine antagonist. The decreased heart rate and blood pressure induced by adenosine was inhibited by cirsimarin. The concentration of cirsimaritin in urine was ~2 μM after administration of 8 mg kg^{-1} cirsimarin. In the kidney and urinary tract the concentration of cirsimaritin produced after cirsimarin ingestion was sufficient to inhibit adenosine receptors and might explain the clinical effectiveness of *Microtea debilis*.

3.2. Chemical Optimization of Affinity of Flavonoids to Adenosine Receptors

The initial finding that flavonoids bind to adenosine receptors (Ji et al., 1995; Ares et al., 1995) has been exploited synthetically to increase receptor selectivity and potency. The structure activity relationships of flavonoids in binding to adenosine receptors were further explored (Karton et al., 1996). The flavone structure may be treated as a molecular template for design of receptor ligands. MRS 1067 (3,6-dichloro-2'-(isopropoxy)-4'-methylflavone; K_i=0.56 μM at human A_3 receptors) was identified through chemical optimization of flavones for receptor binding. MRS 1067 is both relatively potent and highly selective (200-fold) for human A_3 *vs* human A_1 receptors. This derivative effectively blocked the effects of an agonist in the inhibition of adenylyl cyclase in Chinese hamster ovary (CHO) cells expressing either cloned rat or human A_3 receptors (Karton et al., 1996; Jacobson, 1997). MRS 1067 antagonized the effects of the A_3 receptor-selective agonist Cl-IB-MECA to induce a rise in intracellular $[Ca^{2+}]$ in RBL-2H3 rat mast cells (Shin et al., 1996). MRS 1067 is approximately 20-fold selective for rat A_3 receptors.

Table 1. Affinities (K_i, µM) of isoflavone derivatives in a radioligand binding assay of rat A_1 receptors

Compound	R_1	R_2	R_3	A_1[a]
Daidzein	H	H	H	23.4±4.7
Genistein	H	H	5-OH	12.6±1.4
Biochanin A	H	CH_3	5-OH	34.5±4.7
Prunetin	CH_3	H	5-OH	>100

[a] Other compounds (R1, R2, R3) showed <10% displacement at 100µM: (CH_3, CH_3, 5-OH), (H, CH_3, H), (CH_3, CH_3, H), (H, H, 6-OH), (H, H, 6-OCH_3), and 7,3',4'-trihydroxyisoflavone; (±)-equol (7,4'-isoflavandiol) displaced 36±2% at 100 µM.

3.3. Computer Modeling of Flavonoids as Adenosine Antagonists

A quantitative study of the structure activity relationships (SAR) for binding of flavonoids to adenosine A_1, A_{2A}, and A_3 receptors has been conducted using comparative molecular field analysis (CoMFA), which addresses the steric and electrostatic requirements for ligand binding. The relatively rigid ring structure of flavonoids compared to other A_3 receptor antagonists, makes them particularly amenable to CoMFA analysis. Correlation coefficients (cross-validated r^2) of 0.605, 0.595, and 0.583 were obtained for the three subtypes, respectively.

Principal conclusions from the CoMFA analysis are:
a) All three CoMFA models have similar average steric and electrostatic contributions, implying that A_1, A_{2A}, and A_3 have the same relative contribution of steric and electrostatic factors inside the binding cavity. However, the specific distribution of steric and electrostatic interactions for each receptor is different.

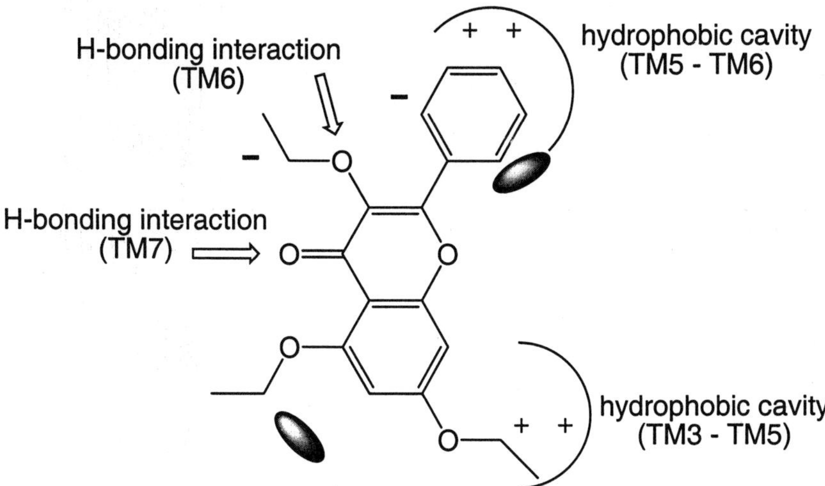

Figure 1. Representation of the steric and electrostatic factors leading to binding of a representative flavone (triethylgalangin) to human A_3 adenosine receptors obtained through CoMFA (Comparative Molecular Field Analysis). + region of positive charge; - region of negative charge; elliptical object = region of tolerance of steric bulk.

b) Similarities were seen in the topology of steric and electrostatic regions with A_1 and A_3 receptors (Fig. 1), but not the A_{2A} receptor. This is in accord with the structure-activity relationship data in which similarity between A_1 and A_3 receptors has been demonstrated.

c) C-2 chromone phenyl ring substituents are considered important for the binding affinity for all adenosine receptors. This phenyl ring may interact in a similar hydrophobic pocket inside the receptor binding-site.

d) An interesting consideration about A_1/A_3 selectivity can be deduced from CoMFA contour map analysis. An important region of favorable steric bulk interaction is located around the 2'-position of the phenyl ring, in the A_3 model. This is consistent with the experimental data for MRS 1067, which is among the most selective flavonoid antagonists of human A_3 receptors.

e) The presence of a C-6 substituent in the chromone moiety is well tolerated, and increases the A_1/A_3 selectivity.

3.4. Identification of Additional Classes of Flavonoids as Adenosine Antagonists

Aurones contain fused 5- and 6-membered rings, and like genistein, naturally occur in soy. We have screeened a number of aurones in binding to adenosine receptors (Table 2) and found considerable affinity. Hispidol was found to be a selective A_1 receptor antagonist, with a K_i value of 350 nM. Antagonism of the A_1 receptor by hispidol was demonstrated in a functional assay consisting of agonist-induced stimulation of binding of $[^{35}S]GTP-\gamma-S$ to the associated G-proteins (Fig. 2).

Table 2. Affinities (K_i, µM) of aurone derivatives in a radioligand binding assay of rat adenosine receptors

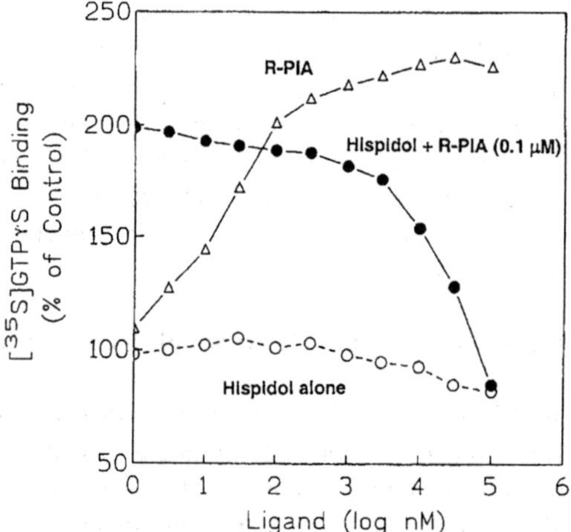

Compound	R_1	R_2	R_3	A_1	A_{2A}
Sulfuretin	H	H	OH	4.44±1.22	28.1±10.1
Aureusidin	OH	H	OH	13.2±4.8	>100
Hispidol	H	H	H	0.352±0.080	52.7±9.1
Maritimetin	H	OH	OH	3.47±0.47	9.35±2.51

The earlier observation that methylation of flavones increased affinities for adenosine receptors led us to also analyze the binding properties of a number of the polymethoxylated flavones in citrus. As shown in Table 3, most polymethoxylated flavones exhibited affinities to the A_1 and A_{2A} receptors in the micromolar range, but showed little selectivity. The several compounds exhibiting greater selectivity had unsubstituted hydroxyl groups at the 3 position. 3-Hydroxy-5,7,3',4'-tetramethoxyflavone had the highest affinity to A_1 (0.3 µM) and a greater than 100-fold selectivity for the A_1 *vs* the A_{2A} subtype. Tight binding to the human A_3 receptors expressed in CHO cells was also observed for sinensetin and 5,7,3',4'-tetramethyl quercetin (K_i=0.34 and 0.31 µM, respectively). The affinity of sinensetin at the human A_3 receptor is the highest exhibited by a naturally occurring flavone thus far examined.

Figure 2. Demonstration of adenosine antagonism by an aurone: Displacement GTP-γ-S binding by hispidol binding to rat A_1 receptors.

Table 3. Affinities (K_i, μM) of Polymethoxylated Flavones in Radioligand Binding Assay to Rat Adenosine Receptors

R5
R4
R6
R3
O
R2
R7
R1
O

Compound	R_1	R_2	R_3	R_4	R_5	R_6	R_7	A_1	A_{2A}
3,5,6,7,8,3',4'-heptamethoxyflavone	OCH_3	OCH_3	OCH_3	OCH_3	OCH_3	OCH_3	OCH_3	15.8±3.4	39.4±5.4
nobiletin (5,6,7,8,3',4'-hexamethoxyflavone)	OCH_3	OCH_3	OCH_3	OCH_3	OCH_3	OCH_3	H	3.74±0.89	21.5±5.3
tangeretin (5,6,7,8,4'-pentamethoxyflavone)	OCH_3	OCH_3	OCH_3	OCH_3	H	OCH_3	H	4.45±0.82	23.7±3.8
tetramethylscutellarein (5,6,7,4'-tetramethoxyflavone)	OCH_3	OCH_3	OCH_3	H	H	OCH_3	H	1.46±0.14	14.3±2.1
sinensetin (5,6,7,3',4'-pentamethoxyflavone)	OCH_3	OCH_3	OCH_3	H	OCH_3	OCH_3	H	0.88±0.13	11.4±0.6
5-desmethylnobiletin 5-hydroxy-6,7,8,3'4'-pentamethoxyflavone	OH	OCH_3	OCH_3	OCH_3	OCH_3	OCH_3	H	4.93±0.14	11.4±3.6
3,5,7,3',4'-pentamethoxyflavone	OCH_3	H	OCH_3	H	OCH_3	OCH_3	OCH_3	17.3±0.4	45.3±6.4
5-hydroxy-3,7,8,3',4'-pentamethoxyflavone	OH	H	OCH_3	OCH_3	OCH_3	OCH_3	OCH_3	10.2±2.0	12.6±6.1
5,8-dihydroxy-3,7,3',4'-tetramethoxyflavone	OH	H	OCH_3	OH	OCH_3	OCH_3	OCH_3	15.4±3.2	>(10^4)
3-hydroxy-5,7,3',4'-tetramethoxyflavone	OCH_3	H	OCH_3	H	OCH_3	OCH_3	OH	0.298±0.03	34.2±6.5

4. DISCUSSION AND CONCLUSIONS

A broad screening of phytochemicals in competitive radioligand binding assays *vs* the high affinity agonist [^{125}I]AB-MECA (N^6-(4-amino-3-iodobenzyl)adenosine-5'-N-methyl-uronamide) has demonstrated that certain naturally occurring flavonoids show affinity at micromolar levels to cloned human brain A$_3$-adenosine receptors (Ji et al., 1995). A later study (Karton et al., 1996) concluded that the affinity of common phytochemicals to adenosine receptors suggests that a wide range of natural substances in the diet may potentially antagonize the effects of endogenous adenosine. Many flavonoids described above appear to be selective antagonists for the A$_1$ or A$_3$ receptors. Thus, adenosine receptor antagonism may be important in the range of biological activities reported for such flavonoids.

We have used CoMFA analysis to distinguish steric and electrostatic requirements for binding of flavones to A$_1$, A$_{2A}$, and A$_3$ receptors. In this manner, the design or discovery of more potent and selective flavonoid antagonists may be accomplished. All three CoMFA models have the same steric and electrostatic contributions, implying similar requirements inside the binding cavity. Similarities were seen in the topology of steric and electrostatic regions with the A$_1$ and A$_3$ receptors, but not the A$_{2A}$. Substitutions on the phenyl ring at the C-2 position of the chromone moiety may be considered important for binding to all adenosine receptors. In the A$_3$ model a region of favorable bulk interaction is located around the 2'-position of the phenyl ring. The presence of a C-6 substituent in the chromone moiety is well tolerated and increases the A$_1$/A$_3$ selectivity. The CoMFA coefficient contour plots provide a self-consistent picture of the main chemical features responsible for the pK$_1$ variations and also result in predictions that agree with experimental values.

Additional SAR studies will be necessary to provide flavonoids that are truly selective adenosine antagonists. Flavonoids are known to bind to a wide range of enzymes (Middleton, 1988; Limasset, 1993; Raghavan, 1995; Dai, 1997), thus the true selectivity of substances that antagonize subtypes of adenosine receptors requires extensive characterization at many binding sites. These non-adenosine-related actions may interfere with their use as pharmacological probes at adenosine receptors. For example, MRS 1067 was found to inhibit (~50% at 28 µM) rat liver cytochrome P450-2C11 (Dai, 1997).

In conclusion, dietary flavonoids make up a significant portion of the human diet. Many of the effects attributed to flavonoids have unexplained mechanisms, but have many similarities to biological effects observed from adenosine receptor antagonism. Through competitive radioligand binding and functional assays, flavonoids were found to be moderately potent antagonists for adenosine receptors, identifying a new class of adenosine receptor ligands, *i.e.* aurones. The aurone hispidol displays selectivity as an A$_1$ adenosine receptor antagonist. Computer modeling of the interactions of flavonoids and A$_1$, A$_{2A}$, and A$_3$ adenosine receptors defined the steric and electronic features of this recognition. These findings suggest antagonism of adenosine receptors, and potentially other G protein-coupled receptors, might be linked to a significant number of the effects proposed for flavonoids.

5. REFERENCES

Ares, J. J., Pong, S. F., Outt, P. E., Blank, M. A., Murray, P. D., and Portlock, D. E., 1995, The binding of flavonoids to adenosine receptors, ACS 209th National Meeting, Chicago, IL, April 1995, Abstract MEDI 190.

Barnes, S., Kim, H., Darley-Usmar, V., Patel, R., Xu, J., Boersma, B., and Luo, M., 2000, Beyond ERa and ERb: Estrogen receptor binding is only part of the isoflavone story, *J. Nutr.* **130**:656S-657S.

Beaven, M. A., Ramkumar, V., and Ali, H., 1994, Adenosine A_3 receptors in mast cells, *Trends Pharmacol. Sci.* **15**:13-14.

Beretz, A., and Cazenave, J. P., 1988, The effect of flavonoids on blood-vessel wall interactions, In *Plant Flavonoids in Biology and Medicine II: Biochemical, Cellular, and Medicinal Properties.* Cody, V. Middleton, E., Jr., Harborne, and J. B., Beretz, A., eds., Alan R. Liss, New York, pp. 187-200.

Coward, L., Barnes, N., Setchell, K., and Barnes, S., 1993, Genistein, daidzein, and their B-glycoside conjugates: antitumor isoflavones in soybean foods from American and Asian diets, *J. Agric. Food Chem.* **41**:1961-1967.

Dai, R., Jacobson, K. A., Robinson, R. C., and Friedman, F. K., 1997, Differential effects of flavonoids on testosterone-metabolizing cytochrome P450s, *Life Sciences. Pharmacology Letters* **61**:PL75-PL80.

Hasrat, J. A., De Bruyne, T., DeBacker, J. P., Vauquelin, G., and Vlietinck, A. J., 1997, Cirsimarin and cirsimaritin, flavonoids of *Microtea debilis* (Phytolaccaceae) with adenosine antagonistic properties in rats: leads for new therapeutics in acute renal failure. *J Pharm. Pharmacol.* **49**:1150-1156.

Jacobson, K. A., Kim, H. O., Siddiqi, S. M., Olah, M. E., Stiles, G., and von Lubitz, D. K. J. E., 1995, A_3 adenosine receptors: Design of selective ligands and therapeutic prospects, *Drugs of the Future* **20**:689-699.

Jacobson, K. A., Park, K. S., Jiang, J. -I., Kim, Y. -C., Olah, M. E., Stiles, G. L., and Ji, X. D., 1997, Pharmacological characterization of novel A_3 adenosine receptor-selective antagonists. *Neuropharmacology* **36**:1157-1165.

Jacobson, K. A., and Suzuki, F., 1997, Recent developments in selective agonists and antagonists acting at purine and pyrimidine receptors, *Drug Devel. Res.* **39**:289-300.

Ji, X. D., Melman, N., and Jacobson, K. A., 1995, Interactions of flavonoids and other phytochemicals with adenosine receptors, *J. Med Chem.* **39**:781-788.

Karton, Y., Jiang, J. L., Ji, X. D., Melman, N., Olah, M. E., Stiles, G. L., and Jacobson, K. A., 1996, Synthesis and biological-activities of flavonoid derivatives as adenosine receptor antagonists, *J. Med. Chem.* **39**:2293-2301.

Limasset, B., le Doucen, C., Dore, J. C., Ojasoo, T., Damon, M., and Crastes de Paulet, A., 1993, Effects of flavonoids on the release of reactive oxygen species by stimulated human neutrophils: Multivariate analysis of structure-activity relationships (SAR), *Biochem. Pharmacol.* **46**:1257-1271.

Manthey, J. A., 2000, Biological properties of flavonoids pertaining to inflammation, *Microcirculation* **7**:S29-S34.

Middleton, E. Jr., 1988, Some biological properties of plant flavonoids, *Ann. Allergy* **61**:53-57.

Middleton, E. Jr., and Ferriola, P., 1988, Effect of flavonoids on protein kinase C: relationship to inhibition of human basophil histamine release, In *Plant Flavonoids in Biology and Medicine II: Biochemical, Cellular, and Medicinal Properties.* Cody, V. Middleton, E., Jr., Harborne, J. B., and Beretz, A., Eds., Alan R. Liss: New York, NY, pp. 251-266.

Moro, S., Van Rhee, A. M., Sanders, L. H., and Jacobson, K. A., 1998, Flavonoid derivatives as adenosine receptor antagonists: A comparison of the hypothetical receptor binding site based on a comparative molecular field analysis model, *J. Med Chem* **41**:46-52.

Müller, C., 1997, A_1 adenosine receptor antagonists, *Exp. Opin. Ther. Patents* **7**:419-440.

Okajima, F., Akbar, M., Abdul Majid, M., Sho, K., Tomura, H., and Kondo, Y., 1994, Genistein, an inhibitor of protein tyrosine kinase, is also a competitive antagonist for P_1-purinergic (adenosine) receptor in FRTL-5 thyroid cells, *Biochem. Biophys. Res. Comm.* **203**:1488-1495.

Raghavan, K., Buolamwini, J. K., Kohn, K. W., and Weinstein, J. N., 1995, Three dimensional quantitative structure-activity relationship (QSAR) of HIV integrase inhibitors: A comparative molecular field analysis (CoMFA) Study, *J. Med. Chem.* **38**:890-897.

Sajjadi, F. G., Takabayashi, K., Foster, A. C., Domingo, R. C., and Firestein, G. S., 1996, Inhibition of TNF-α expression by adenosine - Role of A_3 adenosine receptors, *J. Immunol.* **156**:3435-3442.

Setchel, K, D., 1996, Overview of isoflavone structure, metabolism, and pharmacokentics, From the *Second International Symposium on the Role of Soy in Preventing and Treating Chronic Disease.*, Brussels, Belgium. September 15-18, 1996.

Shin, Y., Daly, J. W., and Jacobson, K. A., 1996, Activation of phosphoinositide breakdown in a rat RBL-2H3 mast cell line by adenosine analogues: Lack of correlation with affinity for A_3-adenosine receptor, *Drug Devel. Res.* **39**:36-46.

Strickler, J., Jacobson, K. A., and Liang, B. T., 1996, Direct preconditioning of cultured chick ventricular myocytes: Novel functions of cardiac adenosine A_{2A} and A_3 receptors, *J. Clin. Invest.* **98**:1773-1779.

von Lubitz, D. K. J. E., Lin, R. C. S., Popik, P., Carter, M. F., and Jacobson, K. A., 1994, Adenosine A_3 receptor stimulation and cerebral ischemia, *Eur. J. Pharmacol.* **263**:59-67.

Wang, H. K., Yang, Z., Natschke, S. L. M., and Lee, K. H., 1998, Recent advances in the discovery and development of flavonoids and their analogues as anti-tumor and anti-HIV agents, In *Advances in Experimental Medicine and Biology: Flavonoids in the Living System*, Manthey, J. A., and Buslig, B. S., eds, Plenum Press, New York, NY pp. 191-227.

REGULATION OF LIPOPROTEIN METABOLISM IN *HepG2* CELLS BY CITRUS FLAVONOIDS

Elzbieta M. Kurowska[1] and John A. Manthey[2]

1. INTRODUCTION

Elevated levels of blood cholesterol are known to be one of the major risk factors associated with coronary heart disease (CHD), the leading cause of death in North America. The association is largely due to the importance of cholesterol, especially low-density lipoprotein (LDL) cholesterol, in the formation and development of atherosclerotic plaque, the underlying pathological condition of CHD. Dietary intervention has been proven to play an important role in prevention and treatment of hypercholesterolemia. Common dietary strategies aimed to lower high blood cholesterol include reduced intake of dietary saturated fat and cholesterol and increased intake of fiber (Connor and Connor, 1998). Recently, many reports have proposed another approach: increased intake of certain food components and food products with cholesterol-lowering potential (Cook and Samman, 1996).

Previous epidemiological studies showed that high dietary intake of fruit and vegetables is associated with reduced risk of coronary heart disease (Bors et al., 1990). The beneficial effects have been postulated to be due to minor components, especially flavonoids (Cook and Samman, 1996). Cardioprotective effects of flavonoids appear to be largely related to their action as antioxidants and as inhibitors of platelet aggregation (Cook and Samman, 1996). However, some dietary flavonoids have also been reported earlier to lower the risk of heart disease by reducing hypercholesterolemia associated with elevation of atherogenic lipoproteins, LDL and/or VLDL (very low density lipoprotein). Flavonoids from soybean, consisting mainly of the isoflavone, genistein, have been shown to decrease blood levels of VLDL + LDL cholesterol in Rhesus monkeys (Anthony et al., 1996) and VLDL but not

[1] KGK Synergize, Inc., Suite 1030, One London Place, 255 Queens Avenue, London, ON N6A 5R8, Canada; [2] U. S. Department of Agriculture, Agricultural Research Service, Citrus and Subtropical Products Laboratory, 600 Avenue S, NW, Winter Haven, FL 33881

Flavonoids in Cell Function, edited by Béla Buslig and John Manthey.
Kluwer Academic/Plenum Publishers, New York, 2002.

LDL cholesterol in rabbits (Kurowska et al., 1994). Other earlier reports demonstrated that in rats, total cholesterol levels were reduced by feeding plant extracts rich in flavonoids (Choi et al., 1991; Monforte et al., 1995; Rajendran et al., 1996; Yotsumoto et al., 1997). Two principal citrus flavonoids, hesperetin from oranges and naringenin from grapefruit, are structurally similar to genistein, which suggests that they could act as cholesterol-lowering agents. In support, earlier studies demonstrated that in rats, blood cholesterol levels were reduced by feeding hesperidin, a hesperetin-7-O-rutinoside abundant in orange juice, or a flavonoid extract from *Prunus davidiana* containing hesperetin 5-O-glucoside (Choi et al., 1991; Monforte et al., 1995). A substantial blood cholesterol reduction was also observed in hypercholesterolemic rats fed a mixture of hesperidin and naringin (a naturally-occurring glycoside of naringenin) (Bok et al., 1999).

In addition to flavanone glycosides, such as hesperidin and naringin, citrus fruit contain polymethoxylated flavones (PMF's), which are found in citrus peel and as minor components in some citrus juices. Some of the major citrus PMF's include tangeretin, nobiletin, sinensetin, heptametoxyflavone and tetra-O-methylscutellarein. Previous studies demonstrated that citrus PMF's exhibit a number of biological activities including anti-cancer activity in mice and in human breast cancer cells (Guthrie and Carroll, 1998a, 1998b) and potential anti-inflammatory activity through their inhibition of tumor necrosis factor-α production in lipopolysaccharide-stimulated human monocytes (Manthey et al., 1999). However, their possible role in regulating cholesterol metabolism has not been investigated. We examined here the cholesterol-lowering properties of natural and synthetic PMF's using human liver cells *HepG2*.

Recent reports have suggested that citrus flavonoid aglycones are absorbed by the gastrointestinal tract unchanged whereas glycosides have to be first hydrolyzed, possibly by intestinal bacteria or by endogenous glycosidases, which liberate the aglycones (Spencer et al., 1999). In humans, free hesperetin and naringenin were detected in blood after oral administration of pure compounds, hesperidin and naringin, or citrus juices (Ameer et al., 1996). It is therefore possible that citrus flavonoids can influence cholesterol metabolism directly in the liver.

2. CHOLESTEROL-LOWERING ACTIVITY OF DIETARY CITRUS JUICES

The effect of dietary orange juice and grapefruit juice on cholesterol metabolism has been investigated in a rabbit model of experimental hypercholesterolemia (Kurowska et al., 2000a). In this study, animals were fed a semipurified, low-fat, cholesterol-free, casein-based diet which produces hypercholesterolemia with a raised LDL fraction, similar to that observed in humans (Kurowska et al., 1989). The control group received water to drink and the experimental groups were given either orange juice or grapefruit juice (reconstituted concentrates made up to two times the regular strength) instead of water. To ensure that the juice consumption did not affect the intake of other food components, diets given to experimental groups were modified to compensate for additional sugar present in the juices and for any changes in the food intake observed during the study. The results showed that after 3 weeks, the animals given citrus juices had significantly lower LDL cholesterol levels than those given water (43% and 32% reduction for orange juice and grapefruit juice, respectively). Also, rabbits given orange juice but not those given grapefruit juice had significantly elevated serum total triacylglycerols (60% increase) (Kurowska et al., 2000a). The observed changes were associated with significant decreases in liver cholesterol esters (42% decrease in each juice group), but not with increases in fecal excretion of cholesterol and bile acids (Kurowska et al., 2000a). The above data suggests that the reduction of LDL cholesterol was unlikely to be due to citrus pectins acting as cholesterol sequestrants in the intestine. Instead, the results allowed speculation that changes in LDL cholesterol and in liver cholesterol esters might be induced by minor juice components, possibly flavonoids, affecting LDL metabolism directly in the liver. This hypothesis is in agreement with reduction of hypercholesterolemia observed in rats fed hesperidin and naringin (Bok et al., 1999).

3. EFFECT OF ORANGE JUICE ON "GOOD CHOLESTEROL"

The modulation of blood lipids by dietary orange juice has also been investigated in moderately hypercholesterolemic human subjects (Kurowska et al., 2000b). In this study, participants incorporated into their diet one, two or three cups (250 mL each) of orange juice per day sequentially, each dose over a period of 4 wks. This was followed by a 5 wk washout period. The results showed that a consumption of 750 mL but not 250 or 500 mL orange juice per day increased HDL (high density lipoprotein or so called "good") cholesterol by 21%. The highest dose of the juice also significantly decreased the LDL/HDL cholesterol ratio by 16% and increased plasma triacylglycerols by 25% but did not lower LDL cholesterol. The observed lack of change in LDL cholesterol could be due to lower amount of the juice or its minor components consumed by participants of the study than by animals (in previous experiments), and/or due to differences between species. The beneficial rise in HDL cholesterol, which was maintained 5 wks after termination of treatment, remains unexplained. The increases in plasma triacylglycerol, which were not clinically significant, corresponded to similar increases observed in rabbits fed orange juice (Kurowska et al., 2000a). The unexpected changes in HDL cholesterol and in plasma triacylglycerol induced by orange juice should be further investigated.

4. CHOLESTEROL-LOWERING EFFECTS OF CITRUS FLAVANONES

The mechanism by which principal citrus flavanones, hesperetin and naringenin, produce cholesterol-lowering responses has been studied in human hepatoma cell line *HepG2* (Borradaile et al., 1999). These neoplastic cells are commonly used as a model of human liver because they can secrete and catabolize lipoproteins similar to LDL (Thrift et al., 1986). The cholesterol-lowering potential of hesperetin and naringenin was evaluated by incubation of cells in the presence vs. absence of each flavanone, and by a subsequent analysis of apolipoprotein B (*apoB*), the structural protein of LDL, in culture medium. The results showed that both hesperetin and naringenin caused a similar, dose-dependent reduction of overall *apoB* secretion. This was consistent with LDL cholesterol-lowering responses observed in rabbits given citrus juices. Thus, the results confirmed the hypothesis that *in vivo*, citrus flavonoids could affect lipoprotein metabolism directly in the liver. Further studies on the mechanism showed that at the highest non-toxic concentrations (60 µg/mL), both flavanones also significantly reduced (by 42%) the incorporation of ^{14}C-acetate into cellular cholesteryl esters without causing significant changes in label incorporation into free cholesterol and triacylglycerols. This suggested that both flavanones could exert their cholesterol-lowering action by interfering with availability of neutral lipids, especially cholesteryl esters, required for the assembly and secretion of lipoproteins. This remains in agreement with a recent observation reported by our collaborators, who demonstrated that naringenin has ability to suppress the activity of acyl CoA:cholesterol-O-acyltransferase (ACAT) in hepatic microsomes (Wilcox et al., 1998) and with reduced *in vitro* activity of ACAT found in liver tissue of rats fed cholesterol-lowering, hesperidin/naringin-supplemented diet (Bok et al., 1999).

5. CHOLESTEROL-LOWERING ACTIVITY OF NATURAL AND SYNTHETIC PMF'S

Five naturally occurring citrus PMF's and three synthetic PMF's were examined for the cholesterol-lowering potential in *HepG2* cells. In these experiments, confluent *HepG2* cells were preincubated for 24 h in a lipoprotein-free medium in which the fetal bovine serum was replaced by albumin to inhibit cell proliferation and to stimulate synthesis of cholesterol-containing lipoproteins. Cells were subsequently incubated in the same medium in the presence or absence of various PMF's at the range of concentrations sustaining 100% cell viability, as determined by 3-(4,5-dimethylthiazol-2-yl)-2,5-diphenyltetrazolium bromide (MTT) viability assay (Hansen et al., 1989). The PMF's were added to the cell culture medium after solubilization in dimethyl sulfoxide (DMSO) and similar concentrations of DMSO were added to control media. At the end of incubation, culture media were collected and concentrations of the LDL structural protein, *apoB*, were measured by ELISA (Borradaile et al., 1999). The amounts of lipoprotein-associated *apoB* in the media (net *apoB* secretion) were determined as described previously (Borradaile et al., 1999), calculated per mg cell protein and expressed as percent of control. For each flavonoid, a dose-response curve was constructed by plotting the percent change of *apoB* in the medium against flavonoid concentration used in the assay. For each compound, the *apoB* IC_{50} concentration was determined as treatment dose of the compound that inhibits the net *apoB* production by 50%.

The results (Table 1) showed that five out of six natural PMF's tested induced a substantial reduction of medium *apoB*. Tangeretin was the most active, reducing medium *apoB* by 86% when added to cells at a concentration of 25 µg/ mL. Nobiletin and 3,5,6,7,8,3',4'-heptamethoxyflavone showed slightly lower activities (81-83% reduction for concentrations 25-50 µg/ mL). The effects of sinensetin and tetra-O-methylscutellarein were similar, although less pronounced. Due to toxicity, tetra-O-methylscutellarein was tested only at concentrations up to 12.5 µg/ mL. 5-desmethylsinensetin induced only a moderate reduction of medium *apoB* (by 64%) at a concentration 50 µg/mL whereas quercetin pentamethyl ether and quercetin-5,7,3',4'-tetramethyl ether had less pronounced effects. The low activity of the synthetic flavonoids could be related to their structure or to their poor uptake by cells.

As demonstrated in Table 2, IC_{50} concentrations of natural PMF's were generally lower than IC_{50}'s of hesperetin and naringenin. For tangeretin, the IC_{50} concentration was 17-20 times lower than for hesperetin and naringenin, respectively. The 5-desmethylsinensetin and the synthetic PMF, quercetin tetramethyl ether, had IC_{50} concentrations comparable to those observed for citrus flavanones, hesperetin and naringenin. These results suggest that

Table 1. Effects of natural and synthetic PMF's on net *apoB* secretion by *HepG2* cells

Compound	flavonoid[a] μg/mL	*apoB* μg/mL	% *apoB* in medium	% *apoB* reduction
Tangeretin (5,6,7,8,4'-pentamethoxyflavone)	0	0.252±0.20	100±4	-
	3.12	0.080±0.02	32±8	68
	6.25	0.055±0.01	22±2	78
	12.5	0.048±0.01	19±2	81
	25.0	0.035±0.01	14±1	86
Nobiletin (5,6,7,8,3',4'-hexamethoxyflavone)	0	0.55±0.21	100±15	-
	6.25	0.37±0.09	67±16	33
	12.5	0.175±0.03	32±6	68
	25.0	0.10±0.01	18±1	82
	50.0	0.095±0.01	17±1	83
3,5,6,7,8,3',4'-heptamethoxyflavone	0	0.250±0.02	100±8	-
	6.25	0.138±0.02	55±5	45
	12.5	0.080±0.02	32±9	68
	25.0	0.048±0.01	19±3	81
	50.0	0.045±0.01	18±2	82
Sinensetin (5,6,7,3',4'-pentamethoxyflavone)	0	0.93±0.07	100±7	-
	6.25	0.70±0.13	75±14	25
	12.5	0.54±0.04	58±4	42
	25.0	0.35±0.04	38±5	62
	50.0	0.26±0.02	28±5	72
Tetra-*O*-methylscutellarein (5,7,8,4'-tetramethoxyflavone)	0	0.430±0.01	100±10	-
	6.25	0.305±0.03	71±7	29
	12.5	0.255±0.06	59±13	41
5-desmethylsinensetin (5-hydroxy-6,7,3',4'-tetramethoxyflavone)	0	1.15±0.29	100±25	-
	25.0	0.84±0.21	73±18	27
	50.0	0.41±0.20	36±1	64
Quercetin 3,7,3',4'-tetramethyl ether[s]	0	1.45±0.17	100±12	-
	25.0	1.05±0.26	72±18	28
	50.0	0.72±0.07	50±5	50
Quercetin 3,5,7,3',4'-pentamethyl ether[s]	0	2.19±0.61	100±28	-
	25.0	1.58±0.22	72±10	28
	50.0	1.43±0.20	65±9	35

[a] concentration in the medium
[s] synthetic

Table 2. *apoB* IC$_{50}$ concentrations of natural and synthetic flavonoids

Compound	*apoB* IC$_{50}$ (μg/mL)
Tangeretin	2.5
Nobiletin	4.9
3,5,6,7,8,3',4'-heptamethoxyflavone	7.8
Sinensetin	17.8
5-desmethylsinensetin	40.5
Quercetin tetramethyl ether[S]	50.0
Hesperetin	43.0
Naringenin	48.5

[S] synthetic

natural PMF's might be more effective as cholesterol-lowering supplements than hesperetin and naringenin.

The pronounced *apoB*-lowering activity observed in *HepG2* cells exposed to tangeretin and other PMF's renders further investigation. Studies are currently being done to evaluate the cholesterol-lowering activity of PMF's in the animal model of hypercholesterolemia and to determine whether the mechanism by which tangeretin and other PMF's modulate cholesterol metabolism in *HepG2* cells differs from that described for hesperetin and naringenin.

6. REFERENCES

Ameer, B., Weintraub, R. A., Johnson, J. V., Yost, R. A., and Rouseff, R. L., 1996, Flavanone absorption after naringenin, hesperidin, and citrus administration, *Clin. Pharmacol. Ther.* **60**:34-40.

Anthony, M. S., Clarkson, T. B., Hughes, C. L. Jr., Morgan, T. M., and Burke, G. L., 1996, Soybean isoflavones improve cardiovascular risk factors without affecting the reproductive system of peripubertal *Rhesus* monkeys, *J. Nutr.* **126**:43-50.

Bok, S.-H., Lee, S.-H., Park, Y.-B., Bae, K.-H., Son, K.-H., Jeong, T.-S., and Choi, M.-S., 1999, Plasma and hepatic cholesterol and hepatic activities of 3-hydroxyl-3-methyl-glutaryl-CoA reductase and acyl CoA:cholesterol transferase are lower in rats fed citrus peel extract or a mixture of citrus bioflavonoids, *J. Nutr..* **129**:1182-1185.

Borradaile, N. M., Carroll, K. K., and Kurowska, E. M., 1999, Regulation of *HepG2* cell apolipoprotein B metabolism by the citrus flavanones hesperetin and naringenin, *Lipids* **34**:591-598.

Bors, W., Heller, W., Michel, C., and Saran, M., 1990, Flavonoids as antioxidants: determination of radical scavenging efficiencies, *Method. Enzymol.* **186**:343-355.

Choi, J. S., Yokozawa, T., and Oura, H., 1991, Antihyperlipidemic effect of flavonoids from *Prunus davidiana*, *J. Nat. Prod.* **54**:218-224.

Connor, S. L., and Connor, W. E., 1998, Pathogenic and protective nutritional factors in coronary heart disease. In: *Current Prospectives on Nutrition and Health*, Carroll, K. K., ed.., McGill-Queen's University Press, Montreal, PQ, pp. 59-100.

Cook, N.C., and Samman, S., 1996, Flavonoids - chemistry, metabolism, cardioprotective effects, and dietary sources, *J. Nutr. Biochem.* **7**:66-76.

Guthrie, N., and Carroll, K. K., 1998a, Inhibition of mammary cancer by citrus flavonoids, In: *Flavonoids in the Living System*, Manthey, J. A., and Buslig, B. S., eds., Plenum Press, New York, pp.227-236.

Guthrie, N., and Carroll, K. K., 1998b, Inhibition of human breast cancer cell growth and metastasis in nude mice by citrus juices and their constituent flavonoids, In: *Biological Oxidants: Molecular Mechanisms and Health Effects*, Packer, L., and Ong, A. S. H., eds., AOCS Press, Champaign, IL, pp.257-264.

Hansen, M. B., Nielsen, S. E., and Berg, K., 1989, Re-examination and further development of a precise and rapid dye method for measuring cell growth/cell kill, *J. Immunol. Meth.* **119**:203-210.

Kurowska, E. M., Hrabek-Smith, J. M., and Carroll, K. K., 1989, Compositional changes in serum lipoproteins during developing hypercholesterolemia induced in rabbits by cholesterol-free semipurified diets, *Atherosclerosis* **78**:159-165.

Kurowska, E. M., Moffatt, M., and Carroll, K. K., 1994, Dietary soybean isoflavones counteract the elevation of VLDL but not LDL cholesterol produced in rabbits by feeding a cholesterol-free, casein diet, *Proc. Can. Fed. Biol. Soc.* **37**:126 (Abstr.).

Kurowska, E. M., Borradaile, N. M., Spence, J. D., and Carroll, K. K., 2000a, Hypocholesterolemic effects of dietary citrus juices in rabbits, *Nutr. Res.* **20**:121-129.

Kurowska, E. M., Spence, J. D., Jordan, J., Wetmore, S., Freeman, D. J., Piche, L. A., and Serratore, P., 2000b, HDL cholesterol-raising effect of orange juice in subjects with hypercholesterolemia, *Am. J. Clin. Nutr.* **72**:1095-1100.

Manthey, J. A., Grohmann, K., Montanari, A., Ash, K., and Manthey, C. L., 1999, Polymethoxylated flavones derived from citrus suppress tumor necrosis factor-α expression by human monocytes, *J. Nat. Prod.* **62**:441-444.

Monforte, M. T., Trovato, A., Kirjavainen, S., Forestieri, A. M., and Galati, E. M., 1995, Biological effects of hesperidin, a citrus flavonoid. (note II): Hypolipidemic activity on experimental hypercholesterolemia in rat, *Farmaco* **50**:595-599.

Rajendran, S., Deepalakshimi, P. D., Parasakthy, K., Devaraj, H., and Devaraj, S. N., 1996, Effect of tincture of *Crataegus* on the LDL-receptor activity of hepatic plasma membrane of rats fed an atherogenic diet, *Atherosclerosis* **123**:235-241.

Spencer, J. P. E., Chowrimootoo, G., Choudhury, R., Debnam, E. S., Srai, S. K., and Rice-Evans, C., 1999, The small intestine can both absorb and glucuronidate luminal flavonoids, *FEBS Letters* **458**:224-230.

Thrift, R. N., Forte, T. M., Cahoon, B. E., and Shore, V. G., 1986, Characterization of lipoprotein produced by the human liver cell line *HepG2*, under defined conditions, *J. Lipid Res.* **27**:236-250.

Wilcox, L. J., Borradaile, N. M., Kurowska, E. M., Telford, D. E., and Huff, M. W., 1998, Naringenin, a citrus flavonoid, markedly decreases *apoB* secretion in *HepG2* cells and inhibits acyl CoA: cholesterol acyltransferase, *Circulation* **98**:I-537 (Abstr.).

Yotsumoto, H., Yanagita, T., Yamamoto, K., Ogawa, Y., Cha, J. Y., and Mori, Y., 1997, Inhibitory effects of Oren-gedoku-to and its components on cholesteryl ester synthesis in cultured human hepatocyte *HepG2* cells: evidence from the cultured *HepG2* cells and in vitro assay of ACAT, *Planta Med.* **63**:141-145.

ANTI-INFLAMMATORY ACTIONS OF A MICRONIZED, PURIFIED FLAVONOID FRACTION IN ISCHEMIA/REPERFUSION

Ronald J. Korthuis[1*] and Dean C. Gute[1]

1. ABSTRACT

It is now recognized that reperfusion after a prolonged period of reduced or absent blood flow, although necessary to salvage ischemic tissue, initiates a complex series of deleterious reactions which ultimately induce the same effects as ischemia *per se*, i.e., cell injury and necrosis. Work conducted over the past 15 years has uncovered the fact that post-ischemic leukocyte infiltration plays a major role in the reperfusion component of ischemia/reperfusion (I/R) injury. This discovery has led to a concerted research effort directed at identifying interventions that prevent post-ischemic leukocyte adhesion and emigration. Recent work indicates that flavonoids are particularly effective anti-inflammatory agents in the setting of I/R. While the mechanisms underlying the powerful protective effects of these compounds is uncertain, a growing body of evidence indicates that flavonoids are potent anti-oxidants that also act to inhibit the activity of key regulatory enzymes involved in the activation of pro-inflammatory signaling cascades. In addition, it appears that these compounds prevent the expression of specific adhesion molecules involved in leukocyte recruitment, observations which provide the molecular basis for the anti-adhesive properties of these compounds.

---

[1] Department of Molecular and Cellular Physiology, Louisiana State University Health Sciences Center, School of Medicine in Shreveport, Shreveport, LA 71130; *Address for correspondence: Ronald J. Korthuis, Ph.D.. Professor and Vice Chair, Department of Molecular & Cellular Physiology, Louisiana State University Health Sciences Center, 1501 Kings Highway. Shreveport, LA 71130; Tel: (318) 675-6028, FAX: (318) 675-4217; e-Mail: rkorth@lsuhsc.edu

Flavonoids in Cell Function, edited by Béla Buslig and John Manthey.
Kluwer Academic/Plenum Publishers, New York, 2002.

2. INTRODUCTION

Reestablishing the blood supply to ischemic tissues is essential to halt the progression of cellular injury associated with decreased oxygen and nutrient delivery and accumulation of deleterious metabolites. As a consequence, minimizing the time to which a tissue is subjected to low blood flow has long been considered as the only important intervention for diminishing the extent of ischernic injury. However, it is now clear that reperfusion of ischemic tissues initiates a complex series of pathologic events that produce the same end result as prolonged hypoxia, i.e., cellular dysfunction and necrosis (Gute et al., 1998; Gute and Korthuis, 1995; Korthuis and Granger, 1995; Panes and Granger, 1998; Rubin et al., 1996). Recognition of the fact that reperfusion can exacerbate the tissue injury induced by ischemia has resulted in an intensive research effort directed at defining the cellular and molecular events that underlie ischemia/reperfusion (I/R) injury. Indeed, work conducted over the past 15 years has led to the development of the concept that oxidant-induced leukocyte/endothelial cell interactions are largely responsible for the microvascular dysfunction induced by reperfusion (Gute et al., 1998; Gute and Korthuis, 1995; Korthuis and Granger, 1995; Panes and Granger, 1998; Rubin et al., 1996). As a consequence of this discovery, much work has been directed at developing new therapeutic strategies aimed at limiting this inflammatory component of I/R injury. One promising avenue relates to the potential use of flavonoids as a means to counteract post-ischemic leukocyte recruitment. The aim of this article is to briefly review the mechanisms whereby infiltrating leukocytes induce tissue injury, the steps involved in the leukocyte recruitment to inflammatory sites, our current understanding of the mechanisms underlying post-ischemic leukocyte adhesion and emigration, and the evidence which indicates that flavonoids may exert powerful protective effects in the setting of I/R.

3. MECHANISMS OF LEUKOCYTE-DEPENDENT TISSUE INJURY

Infiltrating leukocytes release a variety of oxidants and hydrolytic enzymes which can damage the entire array of biomolecules found in tissues, including structural, contractile, and transport proteins, enzymes, receptors, membrane lipids, glycosaminoglycans, and nucleic acids (Korthuis et al., 1992; Halliwell and Gutteridge, 1989). The molecular pathologic effects of oxidant formation include DNA nicking and other abnormalities, peroxidation of membrane lipid components, and cross-linking and degradation of proteins. These leukocyte-induced alterations in the structure and organization of biomolecules can decimate cellular function. For example, oxidant- and/or protease-dependent alterations in protein structure can modify the activity of critical cellular enzymes. Membrane fluidity and compartmentalization can also be modified by leukocyte-mediated lipid peroxidation. As a consequence, the mobility of plasmalemma receptors and the functions of membrane-linked second messenger systems are disrupted. Leakage of intracellular compounds (for example, lactate dehydrogenase and creatine kinase) into the plasma also occurs as a result of membrane lipid peroxidation. Activated leukocytes also disrupt cell-matrix interactions as a consequence of their focused attack on the glycosaminoglycans and proteins in the substrata supporting cells. This effect may contribute to the endothelial denudation and exfoliation of intestinal epithelial cells noted in severe ischemia/reperfusion. Oxidant- and protease-mediated damage to structural, transport, and contractile proteins and inactivation of receptors and essential enzymes in post-ischemic myocardium and skeletal muscle may explain the contractile dysfunction noted in these tissues during reperfusion.

In addition to these direct consequences of leukocyte activation in I/R, white blood cells also contribute to post-ischemic tissue injury in some organs by a more indirect mechanism (Gute and Korthuis, 1995; Jerome et al., 1994). This is related to the neutrophil-dependent increase in vascular permeability that is induced by I/R. As a consequence of the increased fluid and protein extravasation that accompanies post-ischemic microvascular barrier disruption, interstitial fluid pressure rises. This reduces vascular transmural pressure which results in narrowing or obliteration of lumenal diameter in some capillaries, producing no-reflow. These effects are most prominent in organs that cannot readily expand their extravascular compartments as edema forms. For example, expansion of the brain is re-stricted by the cranial vault, whereas distension of the kidney is limited by its capsular investment. Many skeletal muscles are surrounded by a tight fascial sheath that prevents tissue swelling. No-reflow also develops in the heart but appears to occur by a different mechanism in that leukocyte-capillary plugging may have a more dominant role (Engler et al., 1983). Granulocyte trapping in the microcirculation is promoted during I/R by a process that involves ischemia-induced endothelial cell and neutrophil swelling and reduced leukocyte deformability (Dahlgren et al., 1983). The net effect of capillary no-reflow is limitation of reperfusion despite resolution of the pathologic event that produced ischemia in the first place.

4. STEPS INVOLVED IN LEUKOCYTE RECRUITMENT

The recruitment of white blood cells to sites of inflammation involves a complex sequence of highly coordinated steps involving adhesion molecules expressed on the surface of both leukocytes and endothelial cells (reviewed in Panes and Granger, 1998; Korthuis and Gute, 1999). The first step in this process, margination, involves the movement of leukocytes from the center stream of the flowing blood to the vascular wall, a process that is dependent on radial dispersive forces that develop as a consequence of the faster moving erythrocytes slipping past the larger and less flexible white blood cells as blood exits capillaries into postcapillary venules. Upon encountering the vascular wall, circulating leukocytes may begin to form weak adhesive interactions (rolling) if the appropriate adhesive ligands are present. Leukocyte rolling is mediated by P- and E-selectin on endothelial cells and the counterligands L-selectin and PSGL-1 on leukocytes. The selectins can also interact with sialyl Lewis X antigens expressed on both cell types. Selectin-dependent leukocyte rolling slows their transit along postcapillary venules and allows them to monitor their local environment for the presence of activating factors released from inflamed tissue. In the presence of such factors, the rolling leukocyte can modulate the surface expression and/or organization (*e.g.,* clustering) of adhesive structures (CD11/CD18) that favor the transition from leukocyte rolling to stationary (firm) adhesion. Statio-nary adhesion is dependent on the formation of interactions between CD11/CD18 on rolling leukocytes and ICAM-1 on endothelial cells. Once the adherent leukocyte arrests on the walls of postcapillary venules, the white cell begins to flatten out, extends pseudopods between endothelial cells, and moves into the extravascular space. By assuming a less cylindrical geometry during the process of flattening out, the shear stress associated

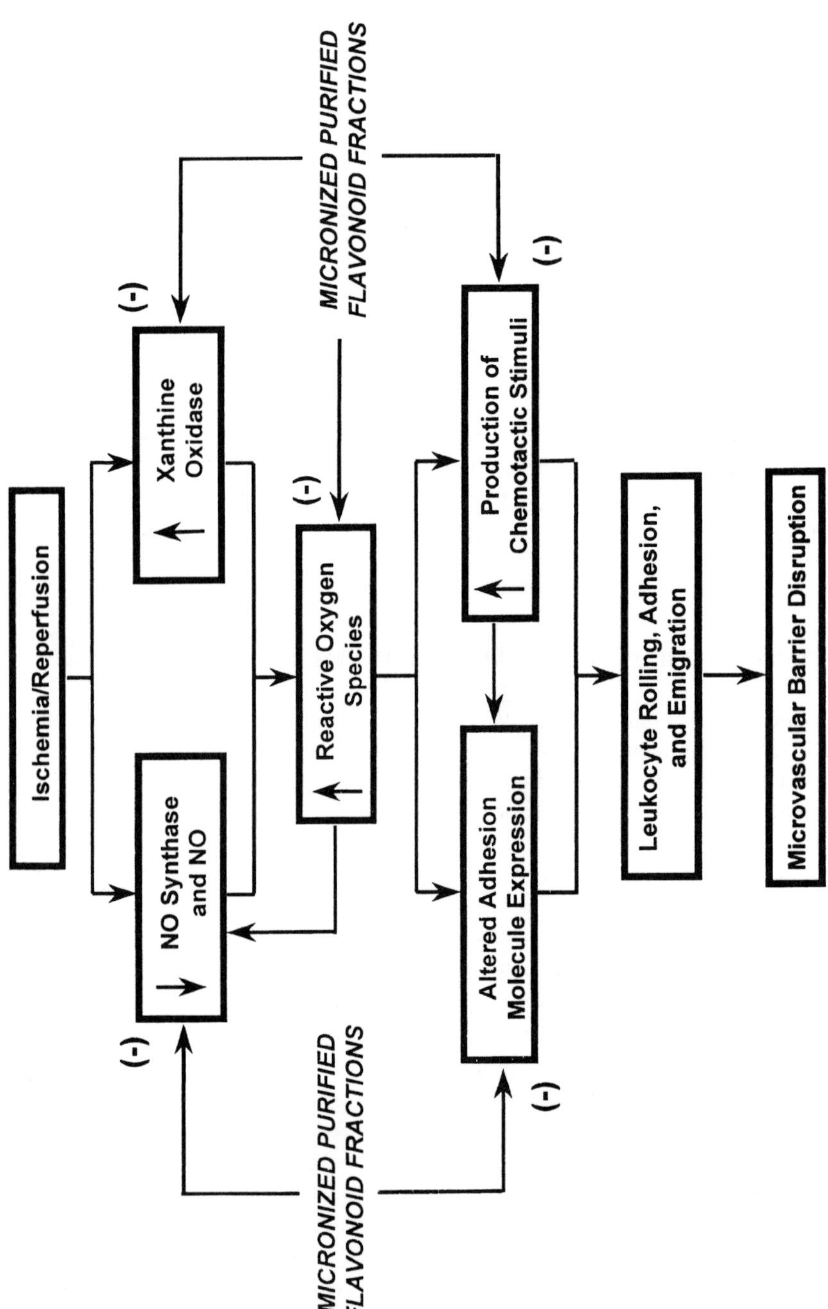

Figure 1. Mechanisms involved in the recruitment of leukocytes during reperfusion of ischemic tissues and potential points in the scheme where flavonoids may exert their anti-inflammatory effects. See text for explanation.

with flowing blood (which tend to sweep leukocytes away from the vascular wall) exert a less prominent force on the adherent cell. The stationary leukocyte uses the gradient in concentration of soluble chemoattractants liberated from the injured tissues as a directional cue to move from the vascular to interstitial compartment, being guided in its transit across the endothelium by interactions with platelet endothelial cell adhesion molecule-1 (PECAM-1).

4. LEUKOCYTE RECRUITMENT TO POST-ISCHEMIC TISSUES

The processes involved in the recruitment of leukocytes to ischemic tissues is complex and multifactorial (see Fig. 1 and references by Gute et al., 1998; Gute and Korthuis, 1995; Korthuis and Granger, 1995; Panes and Granger, 1998; Rubin et al., 1996). Reactive oxygen species generated by xanthine oxidase and other enzymes (*e.g.,* NAD(P)H oxidase) promote the formation of proinflammatory stimuli (*e.g.,* platelet-activating factor, leukotriene B_4 (LTB_4) , and activated complement components), modify the expression of adhesion molecules on the surface of leukocytes and endothelial cells (*e.g.,* CD11b/CD18, P-selectin and intercellular adhesion molecule-1), and reduce levels of the potent anti-adhesive agent nitric oxide. This latter effect is exacerbated by a post-ischemic decline in tissue nitric oxide synthase activity. The resulting imbalance between the production of reactive oxygen species and NO may serve to amplify the intense inflammatory responses elicited by I/R. Coincident with these changes, perivascular cells (*e.g.,* macrophages, mast cells) become activated and release other inflammatory mediators (*e.g.,* histamine, cytokines, LTB_4). As a consequence of these events, leukocytes begin to form adhesive inter-actions with postcapillary venular endothelium. More recent work indicates that platelets play an important role in the adhesion of leukocytes to the post-ischemic microvasculature, possibly by binding to endothelial cells and providing a P-selectin rich platform onto which leukocytes can roll and adhere.

The activated leukocytes emigrate into the tissues, inducing microvascular barrier dysfunction via release of oxidants and hydrolytic enzymes. Thus, enhanced vascular protein leakage is one of the earliest signs of microvascular dysfunction elicited by I/R. In addition to this direct cellular injury induced by leukocytes, emigrating white blood cells also contribute to nutritive perfusion failure (fewer perfused capillaries, i.e., capillary no-reflow) in post-ischemic tissues, as outlined above (Gute et al., 1998; Gute and Korthuis, 1995; Korthuis and Granger, 1995; Panes and Granger, 1998; Rubin et al., 1996).

5. FLAVONOIDS EXERT POWERFUL ANTI-INFLAMMATORY ACTIONS IN I/R

Recognition of the fact that activated neutrophils are largely responsible for the production of microvascular dysfunction induced by I/R has led to a considerable research effort directed at evaluating the potential for inhibition of leukocyte adherence to postcapillary venular endothelium as a novel approach to the treatment of reperfusion injury. Since a number of different flavonoids have been shown to exert a variety of anti-inflammatory actions including inhibition of inflammatory mediator release, altered function of regulatory enzymes, and reduced formation of reactive oxygen species (Lerond, 1994; Freisenecker et al., 1995; Bouskela et al., 1995; Korthuis and Gute, 1997; Manthey, 2000). Daflon 500 mg, is a purified, micronized flavonoid fraction consisting of 90% diosmin and 10% hesperidin and has been shown exert anti-inflammatory effects in a variety of models. These actions suggest that Daflon 500 mg might be useful in the treatment of disorders characterized by ischemia and reperfusion (I/R). To determine whether Daflon 500 mg might exert anti-inflammatory effects in I/R, we used intravital microscopic approaches to quantify leuko-cyte/endothelial cell adhesive interactions and microvascular permeability changes in single postcapillary venules of cremaster muscles and mesenteries subjected to 60 or 20 min of ischemia, respectively, followed by 60 min of reperfusion in the absence (vehicle administration only) and presence of Daflon 500 mg treatment (Korthuis and Gute, 1999).

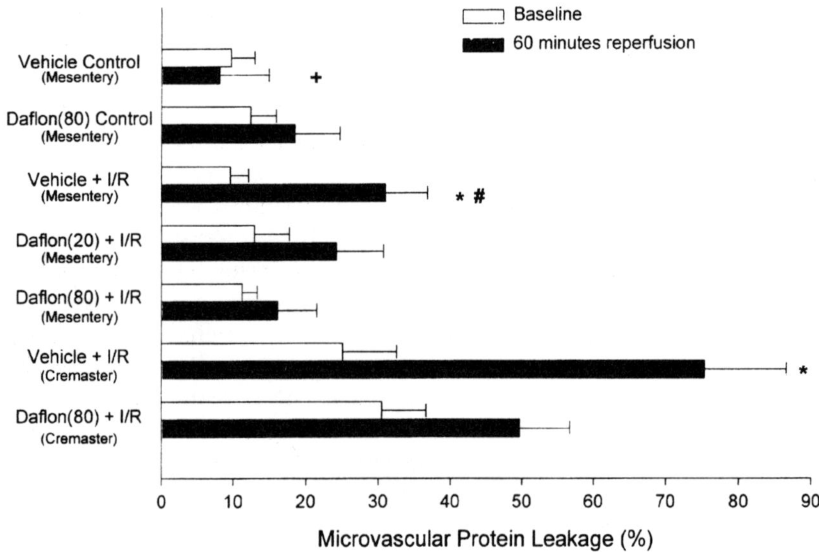

Figure 2. Effects of treatment with a micronized, purified flavonoid fraction (MPFF, Daflon 500 mg) on ischemia/reperfusion (I/R)-induced microvascular protein leakage in the rat cremaster muscle and mesentery. Numbers in parentheses (20 and 80) refer to the doses of MPFF (20 and 80 mg/day). * indicates values different from baseline, # indicates values different from control, and + indicates values different from vehicle + I/R at p<0.05. Data modified from Korthuis and Gute (1999), with permission.

Using these models, we were able to demonstrate that I/R was associated with a marked increase in oxidant production, microvascular barrier dysfunction (as evidenced by increased albumin leakage), and enhanced leukocyte rolling, stationary adhesion, and emigration in both tissues (Figs. 2-4). These proinflammatory effects of I/R were markedly reduced by prior oral administration of Daflon 500 mg (Figs. 2-4). These studies suggest that the beneficial actions of Daflon 500 mg in I/R may be related to its ability to attenuate the establishment of adhesive interactions between circulating leukocytes and postcapillary venular endothelial cells. In the hamster dorsal skin fold preparation, treatment with this flavonoid was shown to reduce macromolecule extravasation but did not affect post-

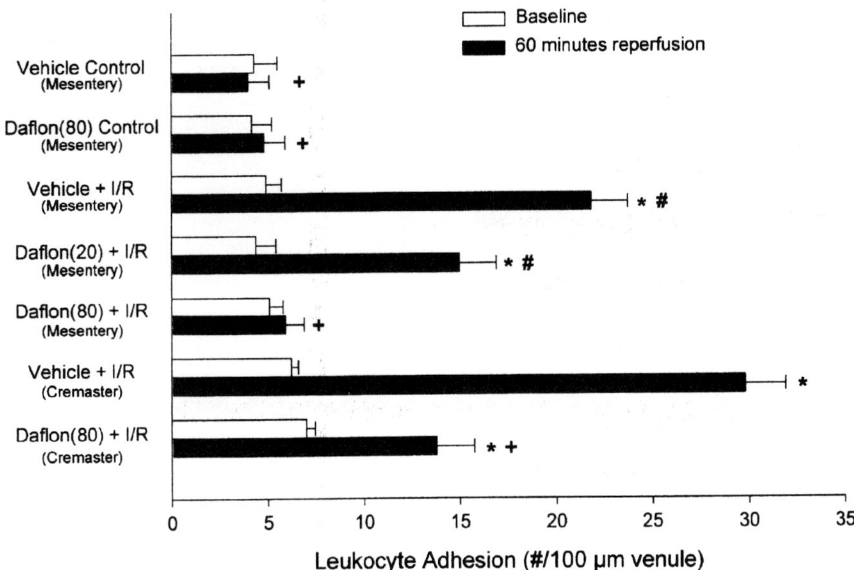

Figure 3. Treatment with a micronized, purified flavonoid fraction (MPFF, Daflon 500 mg) prevents ischemia/reperfusion (I/R)-induced leukocyte adhesion in the rat cremaster muscle and mesentery. Numbers in parentheses (20 and 80) refer to the doses of MPFF (20 and 80 mg/day). * indicates values different from baseline, # indicates values different from control, and + indicates values different from vehicle + I/R at P<0.05. Data modified from Korthuis and Gute(1999), with permission.

ischemic leukocyte adhesion induced by I/R (Nolte et al., 1999). Daflon 500 mg has also been shown to reduce macromolecule leakage in hamster cheek pouches exposed to pro-inflammatory stimuli such as histamine, bradykinin, or leukotriene B, (Bouskela et al., 1995). More recently, Schmid-Schonbein and coworkers (Takase et al., 1999) have shown that this flavonoid prevents leukocyte rolling, adhesion and emigration and reduced paren-chymal cell death in a model thought to mimic the pathologic events (arterial occlusion and reperfusion coincident with increased microvascular pressure) that precipitate the develop-ment of chronic venous insufficiency. Indirect contrast to our findings, the post-ischemic increase in oxidant production induced by I/R was not affected by treatment with this flavonoid fraction in this study (Takase et al., 1999). While the reasons for these discrepant findings are not clear, it is possible that the concomitant elevation in microvascular hydro-static pressure in the latter study may induce a more powerful pro-oxidative stress than I/R alone, perhaps as a result of the attendant mechanical strain on microvessels. On the other hand, Bouskela and coworkers (Bouskela et al., 1999) have shown that Daflon 500 mg reduced stationary leukocyte adhesion after an oxidant challenge, an effect similar to that achieved by administration of the anti-oxidant, α-tocopherol.

Although the aforementioned studies clearly establish the efficacy of Daflon 500 mg treatment in reducing leukocyte/endothelial cell adhesive interactions induced by I/R and other inflammatory stimuli, the molecular mechanisms underlying these anti-inflammatory

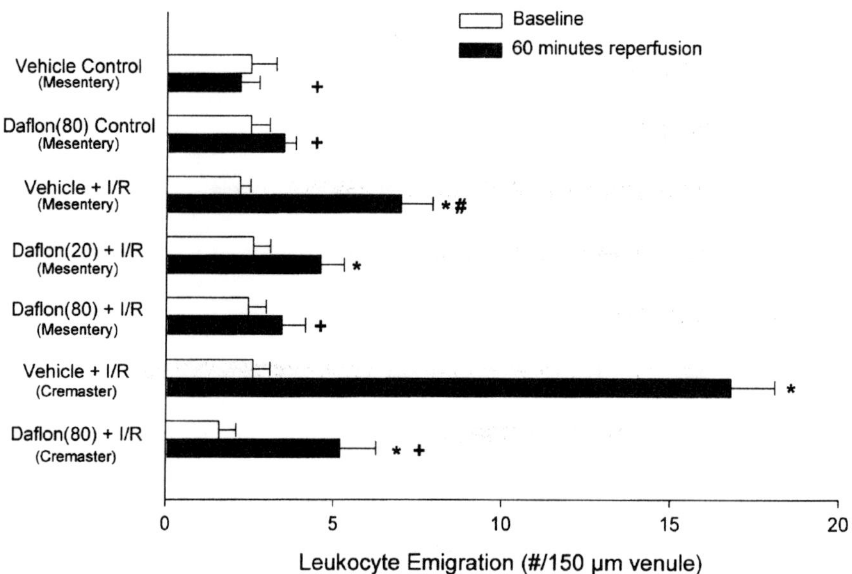

Figure 4. Effects of treatment with a micronized, purified flavonoid fraction (MPFF, Daflon 500 mg) on ischemia/reperfusion (I/R)-induced leukocyte emigration in the rat cremaster muscle and mesentery. Numbers in parentheses (20 and 80) refer to the doses of MPFF (20 and 80 mg/day). * indicates values different from baseline, # indicates values different from control, and + indicates values different from vehicle + I/R at $p < 0.05$. Data modified from Korthuis and Gute (1999), with permission.

actions are not clear. When coupled with the fact that post-ischemic oxidant production plays a central role in reperfusion-induced adhesion molecule expression (Fig. 1), our observation that Daflon 500 mg prevents the formation of reactive oxygen metabolites suggests that the molecular basis for the anti-adhesive effects of this flavonoid may be related to an effect on adhesion molecule expression. Given the marked reductions in post-ischemic leukocyte rolling and adhesion we noted in Daflon 500 mg treated animals, we hypothesized that this flavonoid may prevent the expression of endothelial P-selectin and/or ICAM-1 (Korthuis and Gute, 1999). We used a dual radiolabelled monoclonal antibody approach to quantify the expression of these adhesive structures in an attempt to address this postulate. Despite the profound effect of Daflon 500 mg treatment on post-ischemic leukocyte rolling, treatment with this flavonoid fraction had no effect on the early mobilization of P-selectin by endothelial cells. While this result was unexpected given the results of our immunoneutralization studies demonstrating that P-selectin is required for post-ischemic leukocyte rolling, it is possible Daflon 500 mg acts to induce L-selectin shedding from leukocytes or may modify the surface expression or organization of sialyl Lewis X, thereby impairing the ability of these cells to establish P-selectin-dependent rolling contacts. Indeed, Coleridge-Smith (2000) has demonstrated that Daflon 500 mg substantially reduces the expression of L-selectin on monocytes and neutrophils without affecting the expression of CD11b/CD18 on these leukocyte subtypes.

While it is clear that much further investigation will be required to establish the mechanisms whereby Daflon 500 mg prevents post-ischemic leukocyte rolling, our results regarding the effect of this flavonoid on the molecular determinants of stationary adhesion are more clear. I/R was associated with increased ICAM-1 expression after 4 hours of reperfusion, an effect that was attenuated by Daflon 500 mg treatment (Korthuis and Gute, 1999). This result suggests that Daflon 500 mg may modify transcriptional or translational events involved in the upregulation of ICAM-1. Although the molecular events targeted by Daflon 500 mg treatment are uncertain, it is possible that the flavonoid modifies the activation and/or translocation of the NF-κβ or the phosphorylation and posttranslational modification steps involved in AP-1 activation, transcriptional activator proteins that modulate the expression of ICAM-1 (Panes and Granger, 1998; Korthuis and Gute, 1999).

6. CONCLUSIONS

Infiltrating leukocytes play a major role in the microvascular dysfunction that is induced by reperfusion after prolonged ischemia. Recognition of this fact has led to a concerted research effort directed at developing interventions to limit the post-ischemic inflammatory response. We and others have shown that a micronized, purified flavonoid fraction consisting of 90% diosmin and 10% hesperidin (Daflon 500 mg) is particularly efficacious in this regard. Although we are just now beginning to understand the mechanisms whereby flavonoids exert their anti-inflammatory actions, it is clear that these compounds are potent anti-oxidants which also act to inhibit key regulatory enzymes involved in the generation of powerful proinflammatory signaling cascades. In addition, flavonoids appear to prevent the expression of specific adhesion molecules involved in leukocyte recruitment, an observation which provides the molecular basis for the anti-adhesive effects of these compounds.

7. ACKNOWLEDGMENTS

This work was supported by grants from the National Institutes of Health (HL-54797 and DK-43785) and the Institut de Recherches Internationales Servier.

8. REFERENCES

Bouskela, E., Cyrino, F. Z. G. A., and Lerond, L. 1999, Leukocyte adhesion after oxidant challenge in the hamster cheek pouch microcirculation, *J. Vasc. Res.* **36**(suppl. 1):11-14.

Bouskela, E., Donyo, K. A., and Verbeuren, T. J., 1995, Effects of Daflon 500 mg on increased microvascular permeability in normal hamsters, *Int. J. Microcirc.* **14**(Suppl. 1):22-26.

Coleridge-Smith, P. D., 2000, Micronized purified flavonoid fraction and the treatment of chronic venous insufficiency: Microcirculatory mechanisms, *Microcirculation* **7**:S35-S40.

Dahlgren, M. D., Peterson, M. A., Engler, R. L., and Schmid-Schonbein, G. W., 1984, Leukocyte rheology in cardiac ischemia, *Kroc. Found. Ser.* **16**:271-283.

Engler, R. L., Schmid-Schonbein, G. W., and Pavelec, R. S., 1983, Leukocyte capillary plugging in myocardial ischemia and reperfusion in the dog, *Am. J. Pathol.* **111**:98-111.

Freisenecker, B., Tsai, A. G., and Intaglietta, M., 1995, Cellular basis of inflammation, edema, and the activity of Daflon 500 mg, *Int. J. Microcirc.* **15**(Suppl. 1):17-21.

Freisnecker, B., Tsai, A. G., Allegra, C., and Intaglietta, M., 1994, Oral administration of purified micronized flavonoid fraction suppresses leukocyte adhesion in ischemia-reperfusion injury: In vivo observations in the hamster skin fold, *Int. J. Microcirc.* **14**:50-55.

Gute, D., and Korthuis, R. J., 1995, Role of leukocyte adherence in reperfusion-induced microvascular dysfunction and tissue injury, In *Physiology and Pathophysiology of Leukocyte Adhesion;* Granger, D. N., and Schmid-Schonbein, G. W., eds., Oxford University Press, New York, NY, pp. 359-380.

Gute, D. C., Ishida, T., Yarimizu, K., and Korthuis. R. J., 1998, Inflammation in ischemia/reperfusion induced skeletal muscle injury: Role of neutrophil adhesion and ischemic preconditioning, *Mol. Cell. Biochem.* **179**:169-187.

Halliwell, B., and Gutteridge, J. M. C., 1989, *Free Radicals in Biology and Medicine,* Clarendon Press, Oxford.

Jerome, S. N., Akimitsu, T., and Korthuis, R. J. 1994, Leukocyte adhesion, edema and post-ischemic capillary no-reflow, *Am. J. Physiol.* **267**:H1329-H1336.

Korthuis, R. J., Carden, D. L., and Granger, D. N., 1992, Cellular dysfunction induced by ischemia/reperfusion: role of reactive oxygen metabolites and granulocytes, In *Biological Consequences of Oxidative Stress: Implications for Cardiovascular Disease and Carcinogenesis* Spatz, L., and Bloom, A. D., eds., Oxford University Press, New York, NY, pp. 50-77.

Korthuis, R. J., and Granger, D. N., 1995, Mechanisms of ischemia/reperfusion injury, *Ann. Rev. Physiol.* **57**:311-332,

Korthuis, R. J., and Gute, D. C., 1997, Post-ischemic leukocyte/endothelial cell interactions and microvascular barrier dysfunction in skeletal muscle: Cellular mechanisms and the effect of Daflon 500 mg, *Int. J. Microcirc.* **19**(Suppl. 1):11-17.

Korthuis, R. J., and Gute, D. C., 1999, Adhesion molecule expression in post-ischemic microvascular dysfunction: Activity of a micronized purified flavonoid fraction, *J. Vasc. Res.* **36**(suppl. 1):15-23.

Lerond, L., 1994, Action of Daflon 500 mg on the principal inflammatory mediators, *Phlebology* **9**(Suppl. 1):34-39..

Manthey, J. A., 2000, Biological properties of flavonoids pertaining to inflammation, *Microcirculation* **7**:S29-S34.

Nolte, D., Pickelmann, S., Mollman, M., Schutze, E., Kubler, W., Leiderer, R., and Messmer, K., 1999, Effects of the phlebotriopic drug Daflon 500 on postischemic microvascular disturbances in striated skin muscle: an intravital microscopic study in the hamster, *J. Lab. Clin. Med.* **134**:526-535.

Panes, J., and Granger, D. N., 1998, Leukocyte-endothelial cell interactions: Molecular mechanisms and implications in gastrointestinal disease, *Gastroenterology* **114**:1066-1090.

Rubin, B. B., Romaschin, A., Walker, P. M., Gute, D. C., and Korthuis, R. J., 1996, Mechanisms of post-ischemic skeletal muscle dysfunction: Intervention strategies, *J. Appl. Physiol.* **80**:369-387.

Takase, S., Delano, TA., Lerond, L., Bergan, J. J., and Schmid-Schonbein, G. W., 1999, Inflammation in chronic venous insufficiency: Is the problem insurmountable? *J. Vasc. Res.* **36**(suppl. 1):3-10.

FLAVONOIDS AND GENE EXPRESSION IN MAMMALIAN CELLS

Shiu-Ming Kuo[1]

1. INTRODUCTION

Flavonoids are found to be widely present in the plant kingdom and possess diverse biological activities. The focus of this review is on their regulation of gene expression in mammalian cells. Although a direct DNA-flavonoid interaction could be demonstrated (Ahmed et al., 1994; Strick et al., 2000), this reaction lacks the DNA sequence specificity that is needed for gene regulation. Indeed, no particular gene regulation was reported as a consequence of the DNA-flavonoid interaction. Instead, flavonoids appear to regulate gene expression through interactions with protein transcription factors as described below. Most of the studies in this area examined dietary flavonoids. Some medicinal and synthetic flavonoid-like compounds have also been developed for potential pharmaceutical applications.

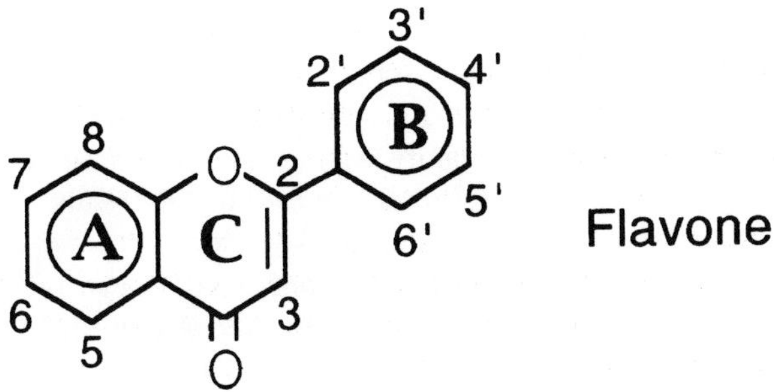

Figure 1. Basic structure of flavonoids. In isoflavones the B-ring is attached at position 3.

[1] Department of Physical Therapy, Exercise and Nutrition Sciences, and Department of Biochemistry, State University of New York at Buffalo, Buffalo, NY 14214

Flavonoids in Cell Function, edited by Béla Buslig and John Manthey.
Kluwer Academic/Plenum Publishers, New York, 2002.

Structurally, flavonoids belong to the family of polyphenolic compounds with the basic structure of flavone as shown in Fig. 1. Subgroups of flavonoids have various modifications including hydroxyl groups attached to different positions on the rings; methylation or glycosylation of the hydroxyl groups; the B ring attached to C3 instead of C2 in isoflavonoids, and the saturation of the C2-3 double bond. As can be predicted from the basic feature, flavonoids are mostly lipid-soluble. Thus, It is not surprising that they diffuse across cell membranes rapidly (Kuo, 1998), and distribute throughout the cells including nuclei (Kuo, 1996). Because of the extensively conjugated double bonds in the basic structure of flavonoids, they are structurally planar providing that the C2-3 bond remains unsaturated as in the case of flavones, flavonols and isoflavones. This particular stereo-structure could further enhance their accessibility to the active site of proteins.

2. INTERACTION WITH STEROID RECEPTORS

The estrogenic activity of flavonoids is well known and the physiological implications of this property have been reviewed (Kurzer and Xu, 1997). At the molecular level, flavonoids increase the expression of endogenous estrogen-responsive genes (Mäkelä et al., 1994; Sathyamoorthy et al., 1994; Hsieh et al., 1998) as well as transfected reporter gene constructs with estrogen-responsive element (ERE) (Miksicek, 1993; Mäkelä et al., 1994; Miksicek, 1994; Kuiper et al., 1998). There is structural specificity for this property. Of the dietary flavonoids tested, isoflavones genistein, daidzein and biochanin A showed the highest stimulatory activity for estrogen-responsive genes. Kaempferol (flavonol), apigenin (flavone) and naringenin (flavanone) were also active but structurally similar flavonol, quercetin, morin and myricetin failed to confer any stimulation (Miksicek, 1993; Mäkelä et al., 1994; Miksicek, 1994). Estrogenic flavonoids bind to estrogen receptors (ER) mimicking the hormone 17ß-estradiol (Miksicek, 1993; Mäkelä et al., 1994; Miksicek, 1994; Kuiper et al., 1998). The ER binding activities of flavonoids were generally lower than that of 17ß-estradiol but flavonoids displayed higher affinity for ERß compared to ERα. In one assay, genistein and 17ß-estradiol were found to have about the same binding activity to ERß (Kuiper et al., 1998). Interestingly, genistein at 1 μM (a concentration close to the physiological range) had higher stimulatory activity than 17ß-estradiol toward ERE-containing reporter gene construct contransfected with ERα or ERß (Kuiper et al., 1998).

Recent work has identified an inhibitory effect of flavonoids, apigenin, quercetin and fisetin, on the expression of vitamin D receptor in human keratinocytes (Segaert et al., 2000). The effect was observed at both protein and mRNA levels and led to a complete suppression of the vitamin D responsiveness in these cells. The significance of this observation has yet to be explored.

3. REGULATION OF ANTIOXIDANT SYSTEM

Flavonoids were shown to have chemical antioxidant activities as reviewed before (Kuo, 1997). Intracellular contents of antioxidant protein thiols were also regulated by some flavonoids. Black tea polyphenols and green tea, but not individual catechin components of the tea, were shown to moderately induce glutathione in normal human Chang liver cells

(Steele et al., 2000). *In vivo*, oral administration of quercitrin also transiently elevated the basal level of glutathione in rat intestine (Galvez et al., 1994). Liver glutathione in mice was depleted by bromobenzene administration and this depletion was partially blocked by the oral administration of a kaempferol glycoside (Sanz et al., 1994). Overall, flavonoids appear to have moderate activity in augmenting cellular glutathione level.

Our lab has investigated the effect of flavonoids on the expression of cysteine-rich metallothionein. Isoflavones, genistein, daidzein and biochanin A, all enhanced the expression of metallothionein in human intestinal cells (Kuo et al., 1998; Kuo and Leavitt, 1999; Kameoka et al., 1999). The effects were time- and dose-dependent. Of the two flavonols tested, kaempferol increased metallothionein levels while quercetin decreased it. Various catechins had little effect. Since the effects observed at the protein level were consistent with that at the mRNA level, the effect could be a result of transcriptional regulation. Besides affecting basal levels of metallothionein in Caco-2 cells, flavonoids also differentially modulated the metallothionein induction by metals (Kuo et al., 1998; Kuo and Leavitt, 1999; Kuo et al., 2001).

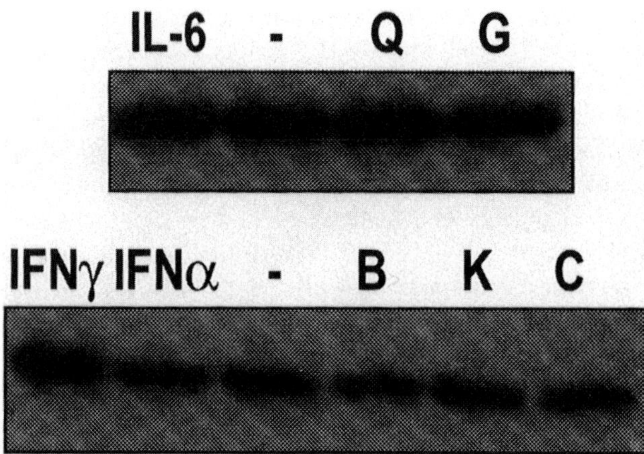

Figure 2. Effect of flavonoids on the level of manganese superoxide dismutase (MnSOD) in Caco-2 cells. Post-confluent Caco-2 cells were treated with 100 µM quercetin (Q), genistein (G), biochanin A (B), kaempferol (K), catechin (C); 100 IU/mL IL-6; 1,000 IU/mL IFN-α or IFN-γ. Cells were harvested 48 hours later and 10 µg protein was loaded in each lane for SDS gel electrophoresis. Western blot was performed with polyclonal antibodies against human MnSOD (Oberley et al., 1989) and detected by ECL plus Western blotting reagent on Hybond C-pure membrane (Amersham Pharmacia Biotech, Piscataway, NJ). Experiment was conducted with triplicate samples and one set is shown. We observed an induction of MnSOD by IFN-γ but IFN-α, IL-6 and flavonoids had no consistent effect.

The effects of flavonoid treatment on mammalian antioxidative enzyme systems were examined. In our studies of cultured intestinal cells, treatment with varieties of flavonoids did not affect the activity of Cu, Zn-superoxide dismutase or catalase as measured by the activity gel assay. The mRNA level of catalase also remained unchanged (Kameoka et al. 1999). Furthermore, the protein level of Mn-superoxide dismutase was not affected by flavonoids in these cells (Fig. 2). Nonetheless, some effects of flavonoids were reported in animal studies. Genistein feeding to SENCAR mice led to small increases in the catalase activity of small intestine and liver. In these mice, superoxide dismutase and glutathione peroxidase activities were also increased in the skin (Cai and Wei, 1996). In a separate study, topical application of a medicinal flavonoid preparation, silymarin, partially prevented the depletion of superoxide dismutase, catalase and glutathione peroxidase activities due to the application of benzoyl peroxide on mouse skin (Zhao et al., 2000). In animal studies as the above two, measurements of the antioxidant protein levels were not available and changes in enzyme activities may not correctly reflect changes in gene expression.

4. REGULATION OF PROLIFERATION-RELATED GENES

Flavonoids are known to inhibit cancer cell growth. At the molecular level, they are generally shown to inhibit the expression of gene products that are linked to cell proliferation while induce the genes that are linked to apoptosis. For example, Ras proteins (p21^{ras}) are G protein-like proteins involved in cell signaling. Mutations at *ras* proto-oncogenes and the subsequent production of overactive Ras proteins have been identified in various human cancers (Bos, 1989). Quercetin treatment can induce cell cycle arrest and inhibit the expression of all three forms of p21^{ras}, K-Ras, H-Ras and N-Ras in human colon cancer cell lines and in primary colorectal tumors (Ranelletti et al., 2000). The expression of another possibly cell cycle-related 17 kDa protein was also inhibited by quercetin in colon cancer cells (Hosokawa et al., 1990). Similarly, the expression of oncogene N-*myc* was decreased by genistein in neuroblastoma cells (Brown et al., 1998), and the expression of c-*myc* was decreased by apigenin in human keratinocytes (Segaert et al., 2000). In breast carcinoma cells, flavopiridol (a synthetic flavonoid) induced cell cycle arrest and down-regulated cyclin D1 which is an important protein for both cell cycle progression and neoplastic transformation (Carlson et al., 1999). The inhibition of cyclin D1 expression was also observed in prostate carcinoma cells treated with a major component of silymarin, silibinin (Zi and Agarwal, 1999), and in rat hepatic stellate cells treated with quercetin (Kawada et al., 1998). Similarly, a reduction of cyclin B1 by genistein treatment was reported in prostate carcinoma cells (Choi et al., 2000). Because of the estrogenic property of genistein as described above, the effect of genistein on cell cycle-related proteins could vary depending on the cell line chosen and the treatment condition. In the absence of serum, genistein at low concentration actually enhanced the synthesis of cyclin D1 in human breast cancer cells mimicking the function of 17ß-estradiol (Dees et al., 1997). The expression of several other tumor markers was decreased by flavonoids but some effects could just be a consequence of cell-cycle blockage. Induction of ornithine decarboxylase by TPA was decreased by green tea as well as its catechin components in 2C5 cells (Steele et al., 2000); and by topical apigenin application to mouse skin (Wei et al., 1990). Prostate-specific antigen synthesis and secretion from prostate carcinoma cells were decreased by silibinin treatment (Zi and Agarwal, 1999). Inhibition of cyclooxygenase-2 activity has been linked to the prevention of colon carcinogenesis. Consistent to the cancer prevention role of flavonoids, the promoter activity of cyclooxygenase-2 in human colon cancer cells was decreased by quercetin and genistein (Mutoh et al., 2000). P21$^{Cip1/WAF1}$ was a known tumor suppressor protein (Sherr, 1993, 1994). The expression of p21$^{Cip1/WAF1}$ was induced by genistein in mouse fibroblast and melanoma cells (Kuzumaki et al., 1998), and in human breast and prostate carcinoma cells (Shao et al., 1998; Choi et al.; 2000). In addition, apigenin and quercetin increased the expression of p21[Cip1/WAF1] in human keratinocytes (Segaert et al., 2000); and silibinin exerted stimulation in human prostate carcinoma cells (Zi and Agarwal, 1999). Besides antiproliferation, genistein also inhibited the invasion of breast cancer cells. The inhibition was associated with modulation of proteins crucial for tumor invasion including a down-regulation of matrix metalloproteinase-9 (MMP-9) and an up-regulation of tissue inhibitor of metalloproteinase (Shao et al., 1998). MMP-9 down regulation was also observed in rabbit synovial fibroblasts and articular chondrocytes by the treatment of various citrus flavonoids (Ishiwa et al., 2000). Using reporter gene assay, it was demonstrated that the inactivation of activator protein-1 (AP-1) site in the MMP-9 promoter could be the molecular mechanism leading to MMP-9 down-regulation (Shao et al., 1998). Genistein appeared to differentially regulate the two transcriptional factors for AP-1 site, c-*Fos* and c-*Jun* (Shao et al., 1998) and thus led to the inactivation of AP-1 site. Under the condition where c-*fos* and c-*jun* expression was induced by ultraviolet light in mouse skin, topical genistein application was shown to decrease the expression of both c-*fos* and c-*jun* (Wang et al., 1998). Analogs of catechin (EGCG, GCG, ECG and CG) also inhibited the activity of AP-1 site in transformed cells probably by down-regulating c-*jun* (Chung et al., 1999).

5. EFFECTS ON DRUG DETOXIFICATION ENZYMES

Enhancement of drug detoxification activity could help the prevention of chemical carcinogenesis. Thus, medicinal chemicals or dietary components that can increase the activity of phase II detoxification enzymes are candidates for cancer chemoprevention. Conversely, modulators of phase I enzymes could give mixed effects. As a whole, flavonoids showed structural-dependent effects on both phase I and phase II enzymes. Tea and its catechin components were shown to be potent inducers of NADPH:quinone reductase and glutathione S-transferase, two phase II enzymes, in human Chang liver cells (Steele et al., 2000). On the other hand, 4'-bromoflavone, a synthetic flavonoid, induced phase I and phase II enzymes concomitantly in rat hepatoma cell culture and in rat tissues, although the effect was more significant for its induction of phase II enzymes (Song et al., 1999). The induction of quinone reductase by 4'-bromoflavone was mediated through the activation of a xenobiotic response element (XRE) and an antioxidant response element (ARE) in the quinone reductase gene (Song et al., 1999). Consistent with this observation, two iso-flavones (equol and genistein) that failed to enhance glutathione S-transferase in a mouse study also could not activate XRE (Helsby et al., 1997). Genistein and equol also did not affect the phase I enzyme activity (Helsby et al., 1997) but quercetin was shown to inhibit cytochrome P-450 1A1 gene expression in Hep G2 cells (Kang et al., 1999).

6. REGULATION OF CYTOKINE RELEASE AND CYTOKINE RESPONSE

Cytokines are essential mediators of cellular responses. Flavonoids, in general, were found to decrease the level of basal and induced secretion of cytokines in varieties of cells.

As summarized below, there are apparently some specificities for cell type, flavonoid structure and cytokine gene. Production of IL-8 by human bronchial gland cells was decreased by genistein dose-dependently (Tabary et al., 1999). In our lab, we found that basal IL-6 secretion from human intestinal Caco-2 cells was decreased by genistein dose-dependently (Kuo et al., 1999). Other flavonoids, including biochanin A, kaempferol, quercetin and EGCG, also decreased the IL-6 secretion although the potency varied (unpublished observation). Genistein also decreased the secretion of IL-2 and leukotriene B4 from lectin-stimulated human mononuclear cells (Atluru et al., 1991). Similarly, lipopolysaccharide (LPS)-induced TNF-α release was decreased by quercetin in the macrophage cell line RAW 264.7 (Wadsworth and Koop, 1999), and by polymethoxylated citrus flavones in human monocytes (Manthey et al., 1999). Although citrus flavones can also decrease the production of macrophage inflammatory protein 1α and interleukin-10 in human monocytes, IL-1ß, IL-6 and IL-8 productions were not affected (Manthey et al., 1999). Stimulated cytokine secretion under other conditions was also affected by flavonoids. Hypoxia-induced transcription of the plasminogen activator inhibitor-1 gene was blocked by genistein in bovine aortic endothelial cells (Uchiyama et al., 2000). Including silymarin in the mouse diet led to a decreased benzoyl peroxide-induced secretion of IL-1α in epidermis (Zhao et al., 2000). Under the stimulated condition, it was proposed that genistein inhibited cytokine secretion through the inhibition of protein tyrosine kinases (Atluru et al., 1991; Uchiyama et al., 2000).

The effect of flavonoids on cytokine secretion is more complicated than just strict inhibition. Transforming growth factor ß (TGFß) is involved in the regulation of cell growth and differentiation. Quercetin was shown to increase TGFß activity in the cultured medium of ovarian cancer cells (Scambia et al., 1994), whereaas genistein only increased TGFβ expression in normal human mammary epithelial cells but not mammary tumor cells (Sathyamoorthy et al. 1998). In a study examining the effects of prepubertal genistein treatment, it was demonstrated that genistein affected the expression of TGFα, epidermal growth factor (EGF) and EGF receptor (EGFR) in the rat mammary gland in an age-dependent fashion (Brown et al., 1998). At 21 days of age, the expressions of TGFα and EGF were not affected by genistein treatment but EGFR level was elevated. In adult rats (at 50 days of age), on the other hand, TGFα level was significantly lower in animals that had prior exposure to genistein but no difference in the level of EGFR was observed.

Nitric oxide production is part of the natural immune response. Consistent with their immunomodulatory property, flavonoids were shown to reduce nitric oxide (NO) production. The observations included inhibition of LPS-inducible nitric oxide synthase (iNOS) by apigenin and kaempferol in macrophages (Liang et al., 1999), decreased NO level by flavone and genistein in LPS-stimulated macrophages (Krol et al., 1995), decreased iNOS activity by various flavonoids in IFN-γ stimulated macrophages (Kobuchi et al., 2000).

Signaling of many cytokines relies on the signal transducer and activator of transcription (STAT) pathway. The binding of cytokine to its receptor is coupled to tyrosine phosphorylation and subsequent nuclear translocation of STAT proteins. STAT proteins then exert their gene regulatory functions by binding to specific sequences in the promoter region of target genes (Hill and Treisman, 1995). Flavonoids, especially genistein, are known inhibitors of protein kinases. Thus, flavonoids could inhibit STAT activation and thus modulate the action of various cytokines. Our preliminary data support an inhibitory effect of genistein and quercetin on the activity of the STAT3 pathway in human intestinal cells. Both basal and IL-6-induced STAT3 activations were decreased. The expression of STAT3 protein was not affected by flavonoids (unpublished observation). Another study has also demonstrated the inhibitory effect of flavonoids on cytokine signaling in non-immune cells. Cytokine-induced adhesion protein expression in endothelial cells was decreased by some flavones (especially apigenin) and flavonols (Gerritsen et al., 1995).

7. SUMMARY

Flavonoids appear to regulate the expression of many genes. As expected, when multiple flavonoids were compared in one study, structure-specificity was always observed. Unfortunately, little information is available regarding the proportion of contribution of various structural elements. Also, we have very limited information on their molecular mechanisms of action. The affinity of flavonoids for ER could explain the stimulatory effect on genes with ERE but other modes of action apparently also exist and need to be further explored.

Physiological relevance is always a concern when investigating the regulation of gene expression by environmental chemicals such as flavonoids. One factor of concern is the *in vivo* concentration of flavonoids. Besides intestinal cells, liver cells and skin cells, other tissues obtain flavonoids through blood circulation. Thus, plasma concentrations of flavonoids are normally discussed. Steady state plasma concentrations of flavonoids are usually not much higher than 1 μM even in populations that consume large amounts of plant material. This concentration is relatively low compared to the concentrations of flavonoids that were commonly used in cell culture systems to demonstrate their effectiveness. Nevertheless, we have evidence that some flavonoids may accumulate in the cell. The effect of quercetin on metallothionein expression in Caco-2 cells persisted for at least 24 hours after its removal from the culture medium (Kuo et al., 1998). Also, long-term treatment of cultured cells with quercetin at low concentrations led to a similar effect on metallothionein expression as one high concentration treatment (Kuo et al., 2001). If intracellular accumulation of certain flavonoids is a shared characteristic for various cell types, it implies that routine ingestion of flavonoids could lead to biological effects at the concentration lower than predicted from a single treatment. Experiments to address possible cell/tissue accumulation of flavonoids are greatly needed.

8. ACKNOWLEDGMENTS

The polyclonal antibodies for manganese superoxide dismutase were a generous gift from Dr. Larry W. Oberley of the University of Iowa. The work was supported partly by NIH grant DK 54728.

9. REFERENCES

Ahmed, M. S., Ramesh, V., Nagaraja, V., Parish, J. H., and Hadi, S. M., 1994, Mode of binding of quercetin to DNA, *Mutagenesis* **9**:193-197.
Atluru, D., Jackson, T. M., and Atluru, S., 1991, Genistein, a selective protein tyrosine kinase inhibitor, inhibits interleukin-2 and leukotriene B4 production from human mononuclear cells, *Clin. Immunol Immunopathol.* **59**:379-387.

Bos, J. L., 1989, *Ras* oncogenes in human cancer: A review, *Cancer Res.* **49**:4682-4689.

Brown, A., Jolly, P., and Wei, H., 1998, Genistein modulates neuroblastoma cell proliferation and differentiation through induction of apoptosis and regulation of tyrosine kinase activity and N-*myc* expression, *Carcinogenesis* **19**:991-997.

Brown, N. M., Wang, J., Cotroneo, M. S., Zhao, Y. -X., and Lamartiniere, C. A., 1998, Prepubertal genistein treatment modulates TGF-α, EGF and EGF-receptor mRNAs and proteins in the rat mammary gland, *Mol. Cell. Endocrinol.* **144**:149-165.

Cai, Q., and Wei, H., 1996, Effect of dietary genistein on antioxidant enzyme activities in SENCAR mice, *Nutr. Cancer* **25**:1-7.

Carlson, B., Lahusen, T., Singh, S., Loaiza-Perez, A., Worland, P. J., Pestell, R., Albanese, C., Sausville, E. A., and Senderowicz, A. M., 1999, Down-regulation of Cyclin D1 by transcriptional repression in MCF-7 human breast carcinoma cells induced by flavopiridol, *Cancer Res.* **59**:4634-4641.

Choi, Y. H., Lee, W. H., and Zhang, L., 2000, p53 independent induction of p21 (WAF1/CIP1), reduction of cyclin B1 and G2/M arrest by the isoflavone genistein in human prostate carcinoma cells, *Jpn. J. Cancer Res.* **91**:164-171.

Chung, J. Y., Huang, C., Meng, X., Dong, Z., and Yang, C. S., 1999, Inhibition of activator protein 1 activity and cell growth by purified green tea and black tea polyphenols in H-*ras*- transformed cells: structure-activity relationship and mechanisms involved, *Cancer Res.* **59**:4610-4617.

Dees, C., Foster, J. S., Ahamed, S., and Wimalasena, J., 1997, Dietary estrogens stimulate human breast cells to enter the cell cycle, *Environ. Health Perspect.* **105**(Suppl 3):633-636.

Galvez, J., De La Cruz, J. P., Zarzuelo, A., De Medina, F. S. Jr., Jimenez, J., and De La Cuesta, F. S., 1994, Oral administration of quercitrin modifies intestinal oxidative status in rats, *Gen. Pharmac.* **25**:1237-1243.

Gerritsen, M. E., Carley, W. W., Ranges, G. E., Shen, C. -P., Phan, S. A., Ligon, G. F., and Perry, C. A., 1995, Flavonoids inhibit cytokine-induced endothelial cell adhesion protein gene expression, *Am. J. Pathol.* **147**:278-292.

Helsby, N. A., Willaims, J., Kerr, D., Gescher, A., and Chipman, J. K., 1997, The isoflavones equol and genistein do not induce xenobiotic-metabolizing enzymes in mouse and in human cells, *Xenobiotica* **6**:587-596.

Hill, C. S., and Treisman, R., 1995, Transcriptional regulation by extracellular signals: mechanisms and specificity, *Cell* **80**:199-211.

Hosokawa, N., Hosokawa, Y., Sakai, T., Yoshida, M., Marui, N., Nishino, H., Kawai, K., and Aoike, A., 1990, Inhibitory effect of quercetin on the synthesis of a possibly cell-cycle-related 17-kDa protein, in human colon cancer cells, *Int. J. Cancer* **45**:1119-1124.

Hsieh, C. -Y., Santell, R. C., Haslam, S. Z., and Helferich, W. G., 1998, Estrogenic effects of genistein on the growth of estrogen receptor-positive human breast cancer (MCF-7) cells *in vitro* and *in vivo*, *Cancer Res.* **58**:3833-3838.

Ishiwa, J., Sato, T., Mimaki, Y., Sashida, Y., Yano, M., and Ito, A., 2000, A citrus flavonoid, nobiletin, suppresses production and gene expression of matrix metalloproteinase 9/gelatinase B in rabbit synovial fibroblasts, *J. Rheumatol.* **27**:20-25.

Kameoka, S., Leavitt, P. S., Chang, C., and Kuo, S. -M., 1999, Expression of antioxidant proteins in human intestinal Caco-2 cells treated with dietary flavonoids, *Cancer Lett.* **146**:161-167.

Kang, Z. C., Tsai, S. J., and Lee, H., 1999, Quercetin inhibits benzo[a]pyrene-induced DNA adducts in human Hep G2 cells by altering cytochrome P-450 1A1 gene expression, *Nutr. Cancer* **35**:175-159.

Kawada, N., Seki, S., Inoue, M., and Kuroki, T., 1998, Effect of antioxidants, resveratrol, quercetin, and N-acetylcysteine, on the functions of cultured rat hepatic stellate cells and Kupffer cells, *Hepatology* **27**:1265-1274.

Kobuchi, H., Virgili, F., and Packer, L., 1999, Assay of inducible form of nitric oxide synthase activity: effect of flavonoids and plant extracts, *Methods in Enzymol.* **301**:504-513.

Krol, W., Czuba, Z. P., Threadgill, M. D., Cunningham, B. D., and Pietsz, G., 1995, Inhibtion of nitric oxide production in murine macrophages by flavones, *Biochem. Pharmacol.* **50**:1031- 1035.

Kuiper, G., Lemmen, J., Carlsson, B., Corton, J., Safe, S., van der Saag, P., van der Burg, B., and Gustafsson, J. -Å., 1998, Interaction of estrogenic chemicals and phytoestrogens with estrogen receptor ß, *Endocrinology* **139**:4252-4263.

Kuo, S.-M., 1996, Antiproliferative potency of structurally distinct dietary flavonoids on human colon cancer cells, *Cancer Lett.* **110**:41-48.

Kuo, S.-M., 1997, Dietary flavonoid and cancer prevention: evidence and potential mechanism, *Critical Rev. Oncogenesis* **8**:47-69.

Kuo, S. -M., 1998, Transepithelial transport and accumulation of flavone in human intestinal Caco-2 cells, *Life Sci.* **63**:2323-2331.

Kuo, S. -M., Leavitt, P. S., and Lin, C. -P., 1998, Dietary flavonoids interact with trace metals and affect metallothionein level in human intestinal cells, *Biol. Trace Element Res.* **62**:135- 153.

Kuo, S. -M., and Leavitt, P. S., 1999, Genistein increases metallothionein expression in human intestinal cells, Caco-2, *Biochem. Cell Biol.* **77**:79-88.

Kuo, S. -M., Leavitt, P. S., and Chang, C., 1999, Cytokines and metallothionein expression in human intestinal Caco-2 cells, *FASEB J.* **13**:A1505.

Kuo, S. -M., Huang, C.-T., Blum, P., and Chang, C., 2001, Quercetin cumulatively enhances copper induction of metallothionein in intestinal cells, *Biol. Trace Element Res.*. (in press).

Kurzer, M., and Xu, X., 1997, Dietary phytoestrogens, *Annu. Rev. Nutr.* **17**:353-381.

Kuzumaki, T., Kobayashi, T., and Ishikawa, K., 1998, Genistein induces p21$^{Cip1/WAF1}$ expression and blocks the G1 to S phase transition in mouse fibroblast and melanoma cells, *Biochem. Biophys. Res. Comm.* **251**:291-295.

Liang, Y. C., Huang, Y. T., Tsai, S. H., Lin-Shiau, S. Y., Chen, C. F., and Lin, J. K., 1999, Suppression of inducible cyclooxygenase and inducible nitric oxide synthase by apigenin and related flavonoids in mouse macrophages, *Carcinogenesis* **20**:1945-1952.

Mäkelä, S., Davis, V. L., Tally W. C., Korkman, J., Salo, L., Vihko, R., Santti, R., and Korach, K. S., 1994, Dietary estrogens act through estrogen receptor-mediated processes and show no antiestrogenicity in cultured breast cancer cells, *Environ. Health Perspect.* **102**:572- 578.

Manthey, J. A., Grohmann, K., Montanari, A., Ash, K., and Manthey, C. L., 1999, Polymethoxylated flavones derived from citrus suppress tumor necrosis factor-alpha expression by human monocytes, *J. Natural Products* **62**:441-444.

Miksicek, R. J., 1993, Commonly occurring plant flavonoids have estrogenic activity, *Mol. Pharmacol.* **44**:37-43.

Miksicek, R. J., 1994, Interaction of naturally occurring non-steroidal estrogens with the recombinant human estrogen receptor, *J. Steroid Biochem. Mol. Biol.* **49**:153-160.

Mutoh, M., Takahashi, M., Fukuda, K., Matsushima-Hibiya, Y., Mutoh, H., Sugimura, T., and Wakabayashi, K., 2000, Suppression of cyclooxygenase-2 promoter-dependent transcriptional activity in colon cancer cells by chemopreventive agents with a resorcin-type structure, *Carcinogenesis* **21**:959-963.

Oberley, L. W., McCormick, M. L., Sierra-Rivera, E., and St. Clair, D. K., 1989, Manganese superoxide dismutase in normal and transformed human embryonic lung fibroblasts, *Free Radical Biol. Med.* **6**:379-384.

Ranelletti, F. O., Maggiano, N., Serra, F. G., Ricci, R., Larocca, L. M., Lanza, P., Scambia, G., Fattorossi, A., Capelli, A., and Piantelli, M., 2000, Quercetin inhibits p21-Ras expression in human colon cancer cell lines and in primary colorectal tumors, *Int. J. Cancer* **85**:438-445.

Sanz, M. J., Ferrandiz, M. L., Cejudo, M., Terencio, M. C., Gil, B., Bustos, G., Ubeda, A., Gunasegaran, R., and Alcaraz, M. J., 1994, Influence of a series of natural flavonoids on free radical generating systems and oxidative stress, *Xenobiotica* **24**:689-699.

Sathyamoorthy, N., Wang, T. T. Y., and Phang, J. M., 1994, Stimulation of pS2 expression by diet-derived compounds, *Cancer Res.* **54**:957-961.

Sathyamoorthy, N., Gilsdorf, J., and Wang, T., 1998, Differential effect of genistein on transforming growth factor β1 expression in normal and malignant mammary epithelial cells, *Anticancer Res.* **18**:2449-2454.

Scambia, G., Panici, P. B., Ranelletti, F. O., Ferrandina, G., De Vincenzo, R., Piantelli, M., Masciullo, V., Bonanno, G., Isola, G., and Mancuso, S., 1994, Quercetin enhances transforming growth factor ß1 secretion by human ovarian cancer cells, *Int. J. Cancer* **57**:211- 215.

Segaert, S., Courtois, S., Garmyn, M., Degreef, H., and Bouillon, R., 2000, The flavonoid apigenin suppresses vitamin D receptor expression and vitamin D responsiveness in normal human keratinocytes, *Biochem. Biophys. Res. Comm.* **268**:237-241.

Shao, Z. -M., Wu, J., Shen, Z. -Z., and Barsky, S. H., 1998, Genistein exerts multiple suppressive effects on human breast carcinoma cells, *Cancer Res.* **58**:1851-1857.

Sherr, C. J., 1993, Mammalian G1 cyclins, *Cell* **73**:1059-1065.

Sherr, C. J., 1994, G1 phase progression: cycling on cue, *Cell* **79**:551-555.

Song, L. L., Kosmeder II, J. W., Lee, S. K., Gerhäuser, C., Lantvit, D., Moon, R. C., Moriarty, R. M., and Pezzuto, J. M., 1999, Cancer chemopreventive activity mediated by 4'-bromoflavone, a potent inducer of phase II detoxification enzymes, *Cancer Res.* **59**:578-585.

Steele, V. E., Kelloff, G. J., Balentine, D., Boone, C. W., Mehta, R., Bagheri, D., Sigman, C. C., Zhu, S., and Sharma, S., 2000, Comparative chemopreventive mechanisms of green tea, black tea and selected polyphenol extracts measured by *in vitro* bioassays, *Carcinogenesis* **21**:63- 67.

Strick, R., Strissel, P. L., Borgers, S., Smith, S. L., and Rowley, J. D., 2000, Dietary bioflavonoids induce cleavage in the *MLL* gene and may contribute to infant leukemia, *Proc. Natl. Acad. Sci.* **97**:4790-4795.

Tabary, O., Escotte, S., Couetil, J. P., Hubert, D., Dusser, D., Puchelle, E., and Jacquot, J., 1999, Genistein inhibits constitutive and inducible NFκβ activation and decreases IL-8 production by human cystic fibrosis bronchial gland cells, *Am. J. Pathol.* **155**:473-481.

Uchiyama, T., Kurabayashi, M., Ohyama, Y., Utsugi, T., Akuzawa, N., Sato, M., Tomono, S., Kawazu, S., and Nagai, R., 2000, Hypoxia induces transcription of the plasminogen activator inhibitor-1 gene through genistein-sensitive tyrosine kinase pathways in vascular endothelial cells, *Arterioscler. Thromb. Vasc. Biol.* **20**:1155-1161.

Wadsworth, T. L., and Koop, D. R., 1999, Effects of the wine polyphenolic quercetin and resveratrol on pro-inflammatory cytokine expression in RAW 264.7 macrophages, *Biochem. Pharmacol.* **57**:941-949.

Wang, Y., E, Y., Zhang, X., Lebwohl, M., DeLeo, V., and Wei, H., 1998, Inhibition of ultraviolet B- induced c-*fos* and c-*jun* expression *in vivo* by a tyrosine kinase inhibitor genistein, *Carcinogenesis* **19**:649-654.

Wei, H., Tye, L., Bresnick, E., and Birt, D. F., 1990, Inhibitory effect of apigenin, a plant flavonoid, on epidermal ornithine decarboxylase and skin tumor promotion in mice, *Cancer Res.* **50**:499-502.

Zi, X., and Agarwal, R., 1999, Silibinin decreases prostate-specific antigen with cell growth inhibition via G1 arrest, leading to differentiation of prostate carcinoma cells: implications for prostate cancer intervention, *Proc. Natl. Acad. Sci.* **96**:7490-7495.

Zhao, J., Lahiri-Chatterjee, M., Sharma, Y., and Agarwal, R., 2000, Inhibitory effect of a flavonoid antioxidant silymarin on benzoyl peroxide-induced tumor promotion, oxidative stress and inflammatory responses in SENCAR mouse skin, *Carcinogenesis* **21**:811-816.